Differential Equations

Mehdi Rahmani-Andebili

Differential Equations

Practice Problems, Methods, and Solutions

 Springer

Mehdi Rahmani-Andebili
Electrical Engineering Department
Montana Technological University
Butte, MT, USA

ISBN 978-3-031-07986-3   ISBN 978-3-031-07984-9   (eBook)
https://doi.org/10.1007/978-3-031-07984-9

This Springer imprint is published by the registered company Springer Nature Switzerland AG
The registered company address is: Gewerbestrasse 11, 6330 Cham, Switzerland

Preface

Differential equations course is one of the main courses of all engineering majors which is taught for freshman or sophomore students. The subjects include different types of first-order differential equations, including separable differential equation, linear differential equation, Bernoulli differential equation, and complete (exact) differential equation; different types of second-order differential equations with constant coefficients, including homogeneous and non-homogeneous differential equations that are solved using inverse operator method, undetermined coefficient method, and Lagrange method; and differential equations with variable coefficients that are solved by using series and Laplace transform. Moreover, many examples about Laplace transform and inverse Laplace transform of different types of functions are presented.

Like the previously published textbooks, the textbook includes very detailed and multiple methods of problem solutions. It can be used as a practicing textbook by students and as a supplementary teaching source by instructors.

To help students study the textbook in the most efficient way, the exercises have been categorized in nine different levels. In this regard, for each problem of the textbook a difficulty level (easy, normal, or hard) and a calculation amount (small, normal, or large) have been assigned. Moreover, in each chapter, problems have been ordered from the easiest problem with the smallest calculations to the most difficult problems with the largest calculations. Therefore, students are suggested to start studying the textbook from the easiest problems and continue practicing until they reach the normal and then the hardest ones. On the other hand, this classification can help instructors choose their desirable problems to conduct a quiz or a test. Moreover, the classification of computation amount can help students manage their time during future exams and instructors give the appropriate problems based on the exam duration.

Since the problems have very detailed solutions and some of them include multiple methods of solution, the textbook can be useful for the under-prepared students. In addition, the textbook is beneficial for knowledgeable students because it includes advanced exercises.

In the preparation of problem solutions, it has been tried to use typical methods to present the textbook as an instructor-recommended one. In other words, the heuristic methods of problem solution have never been used as the first method of problem solution. By considering this key point, the textbook will be in the direction of instructors' lectures, and the instructors will not see any untaught problem solutions in their students' answer sheets.

The Iranian University Entrance Exams for the master's and PhD programs in electrical engineering major is the main reference of the textbook; however, all the problem solutions have been provided by me. The Iranian University Entrance Exam is one of the most competitive university entrance exams in the world that allows only 10% of the applicants to get into prestigious and tuition-free Iranian universities.

Butte, MT, USA   Mehdi Rahmani-Andebili

The Other Works Published by the Author

The author has already published the books and textbooks below with Springer Nature.

Textbooks

Feedback Control Systems Analysis and Design- Practice Problems, Methods, and Solutions, *Springer Nature*, 2022.

Power System Analysis – Practice Problems, Methods, and Solutions, *Springer Nature*, 2022.

Advanced Electrical Circuit Analysis – Practice Problems, Methods, and Solutions, *Springer Nature*, 2022.

AC Electrical Circuit Analysis – Practice Problems, Methods, and Solutions, *Springer Nature*, 2021.

Calculus – Practice Problems, Methods, and Solutions, *Springer Nature*, 2021.

Precalculus – Practice Problems, Methods, and Solutions, *Springer Nature*, 2021.

DC Electrical Circuit Analysis – Practice Problems, Methods, and Solutions, *Springer Nature*, 2020.

Books

Applications of Artificial Intelligence in Planning and Operation of Smart Grid, *Springer Nature*, 2022.

Design, Control, and Operation of Microgrids in Smart Grids, *Springer Nature*, 2021.

Applications of Fuzzy Logic in Planning and Operation of Smart Grids, *Springer Nature*, 2021.

Operation of Smart Homes, *Springer Nature*, 2021.

Planning and Operation of Plug-in Electric Vehicles: Technical, Geographical, and Social Aspects, *Springer Nature*, 2019.

Contents

About the Author

Mehdi Rahmani-Andebili is an assistant professor in the Electrical Engineering Department at Montana Technological University, MT, USA. Before that, he was also an assistant professor in the Engineering Technology Department at State University of New York, Buffalo State, NY, USA, during 2019–2021. He received his first MSc and PhD degrees in electrical engineering (power system) from Tarbiat Modares University and Clemson University in 2011 and 2016, respectively, and his second MSc degree in physics and astronomy from the University of Alabama in Huntsville in 2019. Moreover, he was a postdoctoral fellow at Sharif University of Technology during 2016–2017. As a professor, he has taught many courses such as Essentials of Electrical Engineering Technology, Electrical Circuits Analysis I, Electrical Circuits Analysis II, Electrical Circuits and Devices, Industrial Electronics, Renewable Distributed Generation and Storage, Feedback Controls, DC and AC Electric Machines, and Power System Analysis. Dr. Rahmani-Andebili has more than 200 single-author and first-author publications including journal papers, conference papers, textbooks, books, and book chapters. He is an *IEEE Senior Member* and the permanent reviewer of many credible journals. His research areas include smart grid, power system operation and planning, integration of renewables and energy storages into power system, energy scheduling and demand-side management, plug-in electric vehicles, distributed generation, and advanced optimization techniques in power system studies.

Problems: First-Order Differential Equations

Abstract

In this chapter, different types of first-order differential equations, including separable differential equation, linear differential equation, Bernoulli differential equation, and complete (exact) differential equation, are studied. In this chapter, the problems are categorized in different levels based on their difficulty levels (easy, normal, and hard) and calculation amounts (small, normal, and large). Additionally, the problems are ordered from the easiest problem with the smallest computations to the most difficult problems with the largest calculations.

1.1. Solve the following differential equation.

$$xy' - 2y = x^2$$

Difficulty level ○ Easy ● Normal ○ Hard
Calculation amount ● Small ○ Normal ○ Large

1) $y = cx^2 \ln x + x^2$
2) $y = x^2 e^x + cx^2$
3) $y = x^2 \ln x + cx^2$
4) $y = cx^2 e^x + x^2$

1.2. Calculate the solution of the differential equation below.

$$xdy + ydx = \sin x \, dx$$

Difficulty level ○ Easy ● Normal ○ Hard
Calculation amount ● Small ○ Normal ○ Large

1) $y = c - \cos x$
2) $y = \frac{1}{x}(c - \cos x)$
3) $y = x(c - \cos x)$
4) None of the above

1.3. What is the general solution of the differential equation below?

$$(2xy + x)y' = y$$

Difficulty level ○ Easy ● Normal ○ Hard
Calculation amount ● Small ○ Normal ○ Large

M. Rahmani-Andebili, *Differential Equations*, https://doi.org/10.1007/978-3-031-07984-9_1

1) $y = ce^{-x^2}$

2) $y^2 e^{2y} = cx$

3) $y = cx^{\frac{1}{2}}e^y$

4) $y = ce^{-x^2}y^2$

1.4. In a chemical reaction, the following differential equation with the primary conditions is given. Determine the value of k.

$$\frac{dy}{dt} = -ky, \quad y(t=0) = y_0, \quad y(t=20) = \frac{1}{2}y_0$$

Difficulty level ○ Easy ● Normal ○ Hard
Calculation amount ● Small ○ Normal ○ Large

1) 0.35

2) 0.035

3) 0.025

4) 0.25

1.5. What is the solution of the differential equation below with the given condition?

$$y' = \frac{xy}{y+1}, \quad y(x) \neq 0, \forall x$$

Difficulty level ○ Easy ● Normal ○ Hard
Calculation amount ● Small ○ Normal ○ Large

1) $2y = cx^2$

2) $2y = c \ln y^2$

3) $2y + \ln (cy)^2 = x^2$

4) $y = ce^{x^2}$

1.6. Calculate the value of $y(x=1)$ in the differential equation below with the given primary condition.

$$(x+y)dx + dy = 0, \quad y(x=0) = 0$$

Difficulty level ○ Easy ● Normal ○ Hard
Calculation amount ● Small ○ Normal ○ Large

1) e^{-1}

2) e

3) $-e$

4) $-e^{-1}$

1.7. Solve the differential equation below.

$$x\frac{dy}{dx} - 3y = x^4$$

Difficulty level ○ Easy ● Normal ○ Hard
Calculation amount ● Small ○ Normal ○ Large

1) $y = x^3(x + c)$

2) $y = x^2(x + c)$

3) $y = x(x + c)$

4) $y = -x^3(x + c)$

1.8. What is the general solution of the differential equation below?

$$\frac{dy}{dx} = \tan x \tan y$$

Difficulty level   ○ Easy   ● Normal   ○ Hard
Calculation amount   ● Small   ○ Normal   ○ Large

1) $\cos y \sin x = c$
2) $\cos y \cos x = c$
3) $\sin y \cos x = c$
4) $\sin y \sin x = c$

1.9. What is the integrating factor of the differential equation below?

$$(2y - 3xy^2)dx - x\,dy = 0$$

Difficulty level   ○ Easy   ● Normal   ○ Hard
Calculation amount   ○ Small   ● Normal   ○ Large

1) $x^2 y$
2) xy^2
3) xy^{-2}
4) $x^2 y^2$

1.10. Solve the differential equation below.

$$\frac{dy}{dx} = 1 + \frac{1}{\sin(x - y + 1)}$$

Difficulty level   ○ Easy   ● Normal   ○ Hard
Calculation amount   ○ Small   ● Normal   ○ Large

1) $y = x + 1 - \cos^{-1}x$
2) $y = -x + 1 + \cos^{-1}x$
3) $\cos(y - x + 1) = 2$
4) $\cos(y - x + 1) = x^2 + 1$

1.11. Determine the integrating factor of the differential equation below.

$$(2y - 6x)dx + (3x - 4x^2y^{-1})dy = 0$$

Difficulty level   ○ Easy   ● Normal   ○ Hard
Calculation amount   ○ Small   ● Normal   ○ Large

1) xy^2
2) $x^2 y$
3) $x^2 y^2$
4) xy

1.12. Solve the following differential equation.

$$\left(4x + xy^2\right)dx + \left(y + x^2y\right)dy = 0$$

Difficulty level   ○ Easy   ● Normal   ○ Hard
Calculation amount   ○ Small   ● Normal   ○ Large

1) $y = \frac{C}{1+x^2} - 4$

2) $y^2 = \frac{C}{1+x^2} - 4$

3) $y^2 = \frac{C}{\sqrt{1+x^2}} - 4$

4) $y = \frac{C}{\sqrt{1+x^2}} - 4$

1.13. Solve the following differential equation.

$$y \ln y \, dx + \left(1 + x^2\right)dy = 0$$

Difficulty level   ○ Easy   ● Normal   ○ Hard
Calculation amount   ○ Small   ● Normal   ○ Large

1) $\ln(c \ln x) = -\tan^{-1}y$

2) $\ln y = \tan^{-1}x$

3) $\ln x = \tan^{-1}y$

4) $\ln(c \ln y) = -\tan^{-1}x$

1.14. What is the solution of the differential equation below with the given primary condition?

$$xy' - y = e^{\frac{1}{x}}, \quad y(x = 1) = e$$

Difficulty level   ○ Easy   ● Normal   ○ Hard
Calculation amount   ○ Small   ● Normal   ○ Large

1) $y = xe^{\frac{1}{x}}$

2) $y = -xe^{\frac{1}{x}} + \frac{2e}{x}$

3) $y = 2xe^{\frac{1}{x}} - e$

4) $y = -xe^{\frac{1}{x}} + 2ex$

1.15. Calculate the solution of the differential equation below with the given primary condition.

$$y' \sin x - y \ln y = 0, \quad y\left(x = \frac{\pi}{2}\right) = e$$

Difficulty level   ○ Easy   ● Normal   ○ Hard
Calculation amount   ○ Small   ● Normal   ○ Large

1) $y = e^{\sin x}$

2) $y = e^{\cos x}$

3) $y = e^{\cot \frac{x}{2}}$

4) $y = e^{\tan \frac{x}{2}}$

1.16. Calculate the value of $y(x = 2)$ if the differential equation with the primary condition below is given.

$$xy' + 2y = -\frac{1}{x}, \quad y(x = 1) = 0$$

Difficulty level   ○ Easy   ● Normal   ○ Hard
Calculation amount   ○ Small   ● Normal   ○ Large

1) $-\frac{5}{2}$
2) $\frac{2}{5}$
3) $-\frac{1}{4}$
4) $\frac{5}{2}$

1.17. Solve the following differential equation.

$$\left(y + x^4\right)dx - x\,dy = 0$$

Difficulty level   ○ Easy   ● Normal   ○ Hard
Calculation amount   ○ Small   ● Normal   ○ Large

1) $x^2 y^3 - y^2 = cx^2$
2) $3\sin x + cxy^2 = y$
3) $y = \frac{x^4}{3} - cx$
4) $x - y^2 = cx^5$

1.18. If the differential equation below has an integrating factor of $x^\alpha y^\beta$, what is the value of α?

$$\left(2ye^{x^2 y^2} + 2y^2\right)dx + \left(2xe^{x^2 y^2} + 3xy\right)dy = 0$$

Difficulty level   ○ Easy   ● Normal   ○ Hard
Calculation amount   ○ Small   ○ Normal   ● Large

1) 2
2) 0.5
3) -1
4) 1

1.19. Calculate the solution of the following differential equation.

$$e^y dx + (xe^y + 2y)dy = 0$$

Difficulty level   ○ Easy   ○ Normal   ● Hard
Calculation amount   ● Small   ○ Normal   ○ Large

1) $xe^y + y^2 + c = 0$
2) $-xe^y + y^2 + c = 0$
3) $xe^{-y} + y^2 + c = 0$
4) $-xe^{-y} + y^2 + c = 0$

1.20. What is the general solution of the differential equation below?

$$(x^2 + xy\sin 2x + y\sin^2 x)dx + x\sin^2 x\, dy = 0$$

Difficulty level   ○ Easy   ○ Normal   ● Hard
Calculation amount   ● Small   ○ Normal   ○ Large

1) $xy\sin^2 x + \frac{1}{3}x^3 = c$
2) $xy^2\sin^2 x + \frac{1}{3}x\cos x = c$
3) $xy^2\sin^2 x + \frac{1}{2}x^2\cos x = c$
4) $x^2y^3\sin^2 x + \frac{1}{3}x^2\sin x = c$

1.21. Solve the differential equation below.

$$x\frac{dy}{dx} + y = xy^3$$

Difficulty level   ○ Easy   ○ Normal   ● Hard
Calculation amount   ○ Small   ● Normal   ○ Large

1) $y^2 = \frac{1}{cx^2 - 2x}$
2) $y^2 = \frac{1}{cx^2 + 2x}$
3) $y^2 = cx^2 + 2x$
4) $y^2 = \frac{1}{cx^2 + x}$

1.22. Calculate the general solution of the following differential equation.

$$(x+y)dy + (x-y)dy = 0$$

Difficulty level   ○ Easy   ○ Normal   ● Hard
Calculation amount   ○ Small   ● Normal   ○ Large

1) $2\tan^{-1}(x^2 + y^2) + \ln\left(\frac{y}{x}\right) = c$
2) $\ln\left(\frac{y}{x}\right) + \tan(x-y) = c$
3) $2\tan^{-1}\left(\frac{y}{x}\right) + \ln(x^2 + y^2) = c$
4) $\tan\left(\frac{y}{x}\right) + \ln(x-y) = c$

1.23. Calculate the value of $y(x=0)$ in the differential equation below with the given primary condition.

$$(\cos x)y' + (\sin x)y = 2x\cos^2 x, \quad y\left(\frac{\pi}{4}\right) = -\frac{15\sqrt{2}\pi^2}{32}$$

Difficulty level   ○ Easy   ○ Normal   ● Hard
Calculation amount   ○ Small   ● Normal   ○ Large

1) π^2
2) π
3) $-\pi$
4) $-\pi^2$

1.24. Solve the differential equation below.

$$x \, dy - y \, dx = xy^2 dx$$

Difficulty level   ○ Easy   ○ Normal   ● Hard
Calculation amount   ○ Small   ● Normal   ○ Large

1) $x^2 + \frac{1}{y} = c$

2) $y + xy^2 = c$

3) $\frac{x^2}{2} + \frac{x}{y} = c$

4) $\frac{x}{2} - \frac{y}{2} = c$

1.25. Calculate the solution of the following differential equation.

$$y' + y = \frac{x}{y}$$

Difficulty level   ○ Easy   ○ Normal   ● Hard
Calculation amount   ○ Small   ● Normal   ○ Large

1) $y = x \ln y + c$

2) $y = -x \ln y + c$

3) $y^2 = x - \frac{1}{2} + ce^{2x}$

4) $y^2 = x - \frac{1}{2} + ce^{-2x}$

1.26. What is the general solution of the differential equation below?

$$xy' + y = x^4 y^3$$

Difficulty level   ○ Easy   ○ Normal   ● Hard
Calculation amount   ○ Small   ● Normal   ○ Large

1) $y^2 = x^2(-x^2 + c)$

2) $y^{-2} = x^2(-x^2 + c)$

3) $y^{-2} = x^2(x^2 + c)$

4) $y^2 = x^2(x^2 + c)$

1.27. Which one of the following choices is a solution of the differential equation below that passes through the point of (0, 1)?

$$y' = xy^2 - y$$

Difficulty level   ○ Easy   ○ Normal   ● Hard
Calculation amount   ○ Small   ● Normal   ○ Large

1) $y = \frac{1}{x+1}$

2) $y = \frac{1}{e^x + x}$

3) $y = \frac{1}{3e^x - x - 1}$

4) $y = \frac{1}{2e^x + x - 2}$

1.28. Solve the differential equation below.

$$y' - x^{-1}y = -x^{-1}y^2$$

Difficulty level   ○ Easy   ○ Normal   ● Hard
Calculation amount   ○ Small   ● Normal   ○ Large

1) $y = \frac{c-x}{x}$

2) $y = \frac{c+x}{x}$

3) $y = \frac{x}{x+c}$

4) $y = \frac{x}{c-x}$

1.29. Solve the differential equation below.

$$y' = \sin(x - y)$$

Difficulty level   ○ Easy   ○ Normal   ● Hard
Calculation amount   ○ Small   ● Normal   ○ Large

1) $\tan(x - y) - \sec(x - y) = x + c$

2) $\tan(x - y) + \sec(x - y) = x + c$

3) $\cot(x - y) + \sec(x - y) = x + c$

4) $\cot(x - y) - \sec(x - y) = x + c$

1.30. Calculate the general solution of the following differential equation.

$$y\,dx + x(x^2y - 1)dy = 0$$

Difficulty level   ○ Easy   ○ Normal   ● Hard
Calculation amount   ○ Small   ● Normal   ○ Large

1) $-\frac{y^2}{2x^2} + \frac{y^2}{2} = c$

2) $-\frac{y^2}{2x^2} + \frac{y^3}{3} = c$

3) $-\frac{y^2}{x^2} + y = c$

4) $-\frac{y^2}{x^2} + \frac{y^3}{3} = c$

Abstract

In this chapter, the problems of the first chapter are fully solved, in detail, step-by-step, and with different methods.

2.1. A differential equation with the form of (1) is called a linear differential equation with respect to y based on x.

$$y'(x) + p(x)y(x) = q(x) \tag{1}$$

The solution of a linear differential equation can be determined as follows.

$$y(x) = \frac{1}{\mu(x)} \{\mu(x)q(x)dx + c\} \tag{2}$$

where

$$\mu(x) = e^{\int p(x)dx} \tag{3}$$

Herein, the problem is a linear differential equation that can be solved as follows.

$$xy' - 2y = x^2 \stackrel{\times \frac{1}{x}}{\Rightarrow} y' - \frac{2}{x}y = x$$

$$\Rightarrow \mu = e^{\int \left(-\frac{2}{x}\right)dx} = e^{-2\ln x} = \frac{1}{x^2}$$

$$y = \frac{1}{\frac{1}{x^2}} \left\{ \int \frac{1}{x^2} \times x \, dx + c \right\} = x^2(\ln x + c) \Rightarrow y = x^2 \ln x + cx^2$$

Choice (3) is the answer.

2.2. **The first method**: In the differential equation of $x \, dy + y \, dx = \sin x \, dx$, the terms in the right-hand and left-hand sides are complete (exact) differential equations. Therefore:

$$x \, dy + y \, dx = d(xy) \tag{1}$$

$$\sin x \, dx = d(-\cos x) \tag{2}$$

$$x \, dy + y \, dx = \sin x \, dx \xrightarrow{(1),(2)} d(xy) = d(-\cos x) \Rightarrow xy = -\cos x + c \Rightarrow y = \frac{1}{x}(-\cos x + c)$$

The second method: The problem can be solved as follows.

$$x \, dy + y \, dx = \sin x \, dx \xrightarrow{\times \frac{1}{x \, dx}} \frac{dy}{dx} + \frac{1}{x} y = \frac{\sin x}{x}$$

Now, the problem is in the form of a linear differential equation that can be solved as follows.

$$\mu(x) = e^{\int \frac{1}{x} dx} = e^{\ln x} = x$$

$$y(x) = \frac{1}{x} \left\{ \int x \left(\frac{\sin x}{x} \right) dx + c \right\} \Rightarrow y = \frac{1}{x}(-\cos x + c)$$

Choice (2) is the answer.

Note: A differential equation with the form of (3) is called a linear differential equation with respect to y based on x.

$$y'(x) + p(x)y(x) = q(x) \tag{3}$$

The solution of a linear differential equation can be determined as follows.

$$y(x) = \frac{1}{\mu(x)} \left\{ \mu(x)q(x)dx + c \right\} \tag{4}$$

where

$$\mu(x) = e^{\int p(x)dx} \tag{5}$$

2.3. The problem is a separable differential equation and can be solved as follows.

$$2(xy + x)y' = y \Rightarrow 2x(y+1)\frac{dy}{dx} = y \Rightarrow \frac{2(y+1)}{y}dy = \frac{dx}{x}$$

$$\xrightarrow{\int} \int \left(2 + \frac{2}{y} \right) dy = \int \frac{dx}{x} \Rightarrow 2y + 2\ln y = \ln x + C \Rightarrow \ln y^2 - \ln x = C - 2y \Rightarrow \ln \frac{y^2}{x} = C - 2y$$

$$\Rightarrow \frac{y^2}{x} = e^{C-2y} = e^C e^{-2y} = ce^{-2y} \Rightarrow y^2 e^{2y} = cx$$

Choice (2) is the answer.

2.4. Based on the information given in the problem, we have:

$$\frac{dy}{dt} = -ky \tag{1}$$

$$y(t = 0) = y_0 \tag{2}$$

$$y(t = 20) = \frac{1}{2} y_0 \tag{3}$$

The problem is a separable differential equation and can be solved as follows.

$$\frac{dy}{dt} = -ky \Rightarrow \frac{dy}{y} = -k\,dt \overset{\int}{\Rightarrow} \ln y = -kt + C \Rightarrow y = e^{-kt+C} = e^C e^{-kt} \Rightarrow y = ce^{-kt} \tag{4}$$

Solving (2) and (4):

$$y_0 = ce^{-k \times 0} \Rightarrow c = y_0 \tag{5}$$

Solving (3), (4), and (5):

$$\frac{1}{2} y_0 = y_0 e^{-k \times 20} \Rightarrow e^{-20k} = \frac{1}{2} \Rightarrow -20k = \ln \frac{1}{2} \Rightarrow 20k = \ln 2 \Rightarrow k = \frac{\ln 2}{20} = 0.035$$

Choice (2) is the answer.

2.5. The problem is a separable differential equation and can be solved as follows.

$$y' = \frac{xy}{y+1} \Rightarrow \frac{dy}{dx} = \frac{xy}{y+1} \Rightarrow \frac{y+1}{y} dy = x\,dx$$

$$\overset{\int}{\Rightarrow} \int \left(1 + \frac{1}{y}\right) dy = \int x\,dx \Rightarrow y + \ln y + \ln c = \frac{1}{2}x^2 \Rightarrow 2y + 2\ln(cy) = x^2$$

$$\Rightarrow 2y + \ln(cy)^2 = x^2$$

Choice (3) is the answer.

2.6. A differential equation with the form of (1) is called a linear differential equation with respect to y based on x.

$$y'(x) + p(x)y(x) = q(x) \tag{1}$$

The solution of a linear differential equation can be determined as follows.

$$y(x) = \frac{1}{\mu(x)} \{\mu(x)q(x)dx + c\} \tag{2}$$

where

$$\mu(x) = e^{\int p(x)dx} \tag{3}$$

Based on the information given in the problem, we have

$$(x + y)dx + dy = 0 \tag{4}$$

$$y(x = 0) = 0 \tag{5}$$

The problem can be solved as follows.

$$(x + y)dx + dy = 0 \Rightarrow x + y + \frac{dy}{dx} = 0 \Rightarrow \frac{dy}{dx} + y = -x$$

Now, the problem is in the form of a linear differential equation that can be solved as follows.

$$\Rightarrow \mu(x) = e^{\int dx} = e^x$$

$$y(x) = \frac{1}{e^x} \left\{ \int e^x \times (-x)dx + c \right\} = e^{-x}(e^x - xe^x + c) \Rightarrow y(x) = 1 - x + ce^{-x} \tag{6}$$

Solving (5) and (6):

$$0 = 1 - 0 + c \Rightarrow c = -1 \tag{7}$$

Solving (6) and (7):

$$y(x) = 1 - x - e^{-x}$$

Therefore:

$$y(1) = -e^{-1}$$

Choice (4) is the answer.

2.7. A differential equation with the form of (1) is called a linear differential equation with respect to y based on x.

$$y'(x) + p(x)y(x) = q(x) \tag{1}$$

The solution of a linear differential equation can be determined as follows.

$$y(x) = \frac{1}{\mu(x)} \{\mu(x)q(x)dx + c\} \tag{2}$$

where

$$\mu(x) = e^{\int p(x)dx} \tag{3}$$

The problem can be solved as follows.

$$x\frac{dy}{dx} - 3y = x^4 \Rightarrow \frac{dy}{dx} - \frac{3}{x}y = x^3$$

Now, the problem is in the form of a linear differential equation that can be solved as follows.

$$\Rightarrow \mu(x) = e^{\int -\frac{3}{x}dx} = e^{-3\ln x} = \frac{1}{x^3}$$

$$y(x) = \frac{1}{\frac{1}{x^3}}\left\{\int \frac{1}{x^3} \times (x^3)dx + c\right\} \Rightarrow y = x^3(x+c)$$

Choice (1) is the answer.

2.8. The problem is a separable differential equation and can be solved as follows.

$$\frac{dy}{dx} = \tan x \tan y \Rightarrow \frac{dy}{\tan y} = \tan x \, dx \Rightarrow \cot y \, dy = \tan x \, dx$$

$$\int$$

$$\Rightarrow \ln \sin y = -\ln \cos x + \ln c \Rightarrow \ln \sin x + \ln \cos x = \ln c \Rightarrow \ln(\sin y \cos x) = \ln c$$

$$\Rightarrow \sin y \cos x = c$$

Choice (3) is the answer.

2.9. Based on the form of the choices, the integrating factor should be in the form of $x^\alpha y^\beta$. The problem can be solved as follows.

$$(2y - 3xy^2)dx - x \, dy = 0 \xrightarrow{\times x^\alpha y^\beta} (2x^\alpha y^{1+\beta} - 3x^{1+\alpha}y^{2+\beta})dx - x^{1+\alpha}y^\beta dy = 0 \quad (1)$$

Now, the differential equation of (1) is a complete (exact) differential equation in the form of $P(x, y)dx + Q(x, y)dy = 0$. Therefore:

$$\Delta = Q_X - P_y = 0$$

$$\Rightarrow -(1+\alpha)x^\alpha y^\beta - (2(1+\beta)x^\alpha y^\beta - 3(2+\beta)x^{1+\alpha}y^{1+\beta}) = 0$$

$$\Rightarrow -(\alpha + 2\beta + 3)x^\alpha y^\beta + 3(2+\beta)x^{1+\alpha}y^{1+\beta} = 0 \Rightarrow \begin{cases} \alpha + 2\beta + 3 = 0 \\ 2 + \beta = 0 \end{cases} \Rightarrow \beta = -2, \alpha = 1$$

$$\Rightarrow \mu(x) = xy^{-2}$$

Choice (3) is the answer.

2.10. Based on the information given in the problem, we have:

$$\frac{dy}{dx} = 1 + \frac{1}{\sin(x - y + 1)} \quad (1)$$

The problem can be solved by defining a new variable as follows.

$$x - y + 1 \triangleq u \quad (2)$$

$$\Rightarrow y' = 1 - u' \quad (3)$$

Solving (1)–(3):

$$1 - u' = 1 + \frac{1}{\sin u} \Rightarrow \frac{du}{dx} = -\frac{1}{\sin u} \Rightarrow -\sin u \, du = dx \Rightarrow \cos u = x + c \Rightarrow u = \cos^{-1}(x + c) \qquad (4)$$

Solving (2) and (4):

$$x - y + 1 = \cos^{-1}(x + c)$$

$$\xRightarrow{c \triangleq 0} y = x + 1 - \cos^{-1} x$$

Choice (1) is the answer.

2.11. Based on the form of the choices, the integrating factor should be in the form of $x^\alpha y^\beta$. The problem can be solved as follows.

$$(2y - 6x)dx + (3x - 4x^2 y^{-1})dy = 0$$

$$\xRightarrow{\times x^\alpha y^\beta} (2x^\alpha y^{\beta+1} - 6x^{\alpha+1} y^\beta)dx + (3x^{\alpha+1} y^\beta - 4x^{\alpha+2} y^{\beta-1})dy = 0 \qquad (1)$$

Now, the differential equation of (1) is a complete (exact) differential equation in the form of $P(x, y)dx + Q(x, y)dy = 0$. Therefore:

$$\Delta = Q_X - P_y = 0$$

$$\Rightarrow 3(\alpha + 1)x^\alpha y^\beta - 4(\alpha + 2)x^{\alpha+1} y^{\beta-1} - (2(\beta + 1)x^\alpha y^\beta - 6\beta x^{\alpha+1} y^{\beta-1}) = 0$$

$$\Rightarrow (3\alpha - 2\beta + 1)x^\alpha y^\beta - (8 + 4\alpha - 6\beta)x^{\alpha+1} y^{\beta-1} = 0$$

$$\Rightarrow \begin{cases} 3\alpha - 2\beta + 1 = 0 \\ 4\alpha - 6\beta + 8 = 0 \end{cases} \Rightarrow \alpha = 1, \beta = 2$$

$$\Rightarrow \mu(x) = xy^2$$

Choice (1) is the answer.

2.12. The problem is a separable differential equation and can be solved as follows.

$$(4x + xy^2)dx + (y + x^2 y)dy = 0 \Rightarrow x(4 + y^2)dx = -y(1 + x^2)dy \Rightarrow \frac{y}{4 + y^2}dy = \frac{-x}{1 + x^2}dx$$

$$\xRightarrow{\int} \frac{1}{2}\ln(4 + y^2) = \frac{-1}{2}\ln(1 + x^2) + \ln C \Rightarrow \ln(4 + y^2) + \ln(1 + x^2) = 2\ln C$$

$$\Rightarrow \ln((4 + y^2)(1 + x^2)) = \ln C^2 = \ln c \Rightarrow (4 + y^2)(1 + x^2) = c \Rightarrow 4 + y^2 = \frac{c}{1 + x^2}$$

$$\Rightarrow y^2 = \frac{c}{1 + x^2} - 4$$

Choice (2) is the answer.

2.13. The problem is a separable differential equation and can be solved as follows.

$$y \ln y \, dx + \left(1 + x^2\right) dy = 0 \Rightarrow y \ln y \, dx = -\left(1 + x^2\right) dy \Rightarrow \frac{dy}{y \ln y} = -\frac{dx}{1 + x^2} \tag{1}$$

Changing the variable:

$$\ln y \triangleq u \tag{2}$$

$$\Rightarrow \frac{dy}{y} = du \tag{3}$$

Solving (1)–(3):

$$\frac{du}{u} = -\frac{dx}{1 + x^2} \xrightarrow{\int} \ln(u) + \ln c = -\tan^{-1}x \tag{4}$$

Solving (2) and (4):

$$\ln(\ln y) + \ln c = -\tan^{-1}x$$

$$\Rightarrow \ln(c \ln y) = -\tan^{-1}x$$

Choice (4) is the answer.

2.14. A differential equation with the form of (1) is called a linear differential equation with respect to y based on x.

$$y'(x) + p(x)y(x) = q(x) \tag{1}$$

The solution of a linear differential equation can be determined as follows.

$$y(x) = \frac{1}{\mu(x)} \left\{ \mu(x)q(x)dx + c \right\} \tag{2}$$

where

$$\mu(x) = e^{\int p(x)dx} \tag{3}$$

Based on the information given in the problem, we have:

$$xy' - y = e^{\frac{1}{x}} \tag{4}$$

$$y(x = 1) = e \tag{5}$$

The problem is a linear differential equation that can be solved as follows.

$$xy' - y = e^{\frac{1}{x}} \Rightarrow y' - \frac{1}{x}y = \frac{1}{x}e^{\frac{1}{x}}$$

$$\Rightarrow \mu(x) = e^{\int -\frac{1}{x}dx} = e^{-\ln x} = \frac{1}{x}$$

$$y(x) = \frac{1}{\frac{1}{x}}\left\{\int \frac{1}{x} \times \frac{1}{x}e^{\frac{1}{x}}dx + c\right\} = x\left(-e^{\frac{1}{x}} + c\right) \tag{6}$$

Solving (5) and (6):

$$e = -e + c \Rightarrow c = 2e \tag{7}$$

Solving (6) and (7):

$$y = x\left(-e^{\frac{1}{x}} + 2e\right) \Rightarrow y = -xe^{\frac{1}{x}} + 2ex$$

Choice (4) is the answer.

2.15. Based on the information given in the problem, we have:

$$y'\sin x - y\ln y = 0, \quad y\left(x = \frac{\pi}{2}\right) = e \tag{1}$$

The problem is a separable differential equation and can be solved as follows.

$$y'\sin x - y\ln y = 0 \Rightarrow \frac{dy}{dx}\sin x = y\ln y \Rightarrow \frac{dy}{y\ln y} = \frac{dx}{\sin x}$$

$$\Rightarrow \int \frac{\frac{1}{y}}{\ln y}dy = \int \frac{dx}{\sin x} \Rightarrow \ln(\ln y) = \ln\left(\tan\frac{x}{2}\right) + \ln c \Rightarrow \ln y = c\tan\frac{x}{2} \tag{2}$$

Solving (1) and (2):

$$\ln e = c\tan\frac{\pi}{4} \Rightarrow c = 1 \tag{3}$$

Solving (2) and (3):

$$\ln y = \tan\frac{x}{2} \Rightarrow y = e^{\tan\frac{x}{2}}$$

Choice (4) is the answer.

2.16. A differential equation with the form of (1) is called a linear differential equation with respect to y based on x.

$$y'(x) + p(x)y(x) = q(x) \tag{1}$$

The solution of a linear differential equation can be determined as follows.

$$y(x) = \frac{1}{\mu(x)}\{\mu(x)q(x)dx + c\} \tag{2}$$

where

$$\mu(x) = e^{\int p(x)dx} \tag{3}$$

Based on the information given in the problem, we have:

$$xy' + 2y = -\frac{1}{x}, \quad y(x=1) = 0 \tag{4}$$

The problem is in the form of a linear differential equation that can be solved as follows.

$$xy' + 2y = -\frac{1}{x} \xrightarrow{\times\frac{1}{x}} y' + \frac{2}{x}y = -\frac{1}{x^2}$$

$$\mu(x) = e^{\int \frac{2}{x}dx} = e^{2\ln x} = x^2$$

$$y(x) = \frac{1}{x^2}\left\{ \int x^2 \times \left(-\frac{1}{x^2}\right)dx + c \right\} = \frac{1}{x^2}(-x + c) = -\frac{1}{x} + \frac{c}{x^2} \tag{5}$$

Solving (4) and (5):

$$0 = -1 + c \Rightarrow c = 1 \tag{6}$$

Solving (5) and (6):

$$y(x) = -\frac{1}{x} + \frac{1}{x^2}$$

Hence:

$$y(2) = -\frac{1}{2} + \frac{1}{4} \Rightarrow y(2) = -\frac{1}{4}$$

Choice (3) is the answer.

2.17. A differential equation with the form of (1) is called a linear differential equation with respect to y based on x.

$$y'(x) + p(x)y(x) = q(x) \tag{1}$$

The solution of a linear differential equation can be determined as follows.

$$y(x) = \frac{1}{\mu(x)}\{\mu(x)q(x)dx + c\} \tag{2}$$

where

$$\mu(x) = e^{\int p(x)dx} \tag{3}$$

The problem can be solved as follows.

$$\left(y + x^4\right)dx - x\,dy = 0 \Rightarrow x\,dy = \left(y + x^4\right)dx \Rightarrow \frac{dy}{dx} = \frac{y + x^4}{x} \Rightarrow \frac{dy}{dx} - \frac{1}{x}y = x^3$$

Now, the problem is in the form of a linear differential equation that can be solved as follows.

$$\mu(x) = e^{\int -\frac{1}{x}dx} = e^{-\ln x} = \frac{1}{x}$$

$$y(x) = \frac{1}{\frac{1}{x}}\left\{\int \frac{1}{x} \times x^3\,dx + C\right\} = x\left(\frac{x^3}{3} + C\right) \Rightarrow y = \frac{x^4}{3} - cx$$

Choice (3) is the answer.

2.18. Based on the form of the choices, the integrating factor should be in the form of $x^\alpha y^\beta$. The problem can be solved as follows.

$$\left(2ye^{x^2y^2} + 2y^2\right)dx + \left(2xe^{x^2y^2} + 3xy\right)dy = 0$$

$$\xrightarrow{\times x^\alpha y^\beta} \left(2x^\alpha y^{1+\alpha}e^{x^2y^2} + 2x^\alpha y^{2+\alpha}\right)dx + \left(2x^{1+\alpha}y^\alpha e^{x^2y^2} + 3x^{1+\alpha}y^{1+\alpha}\right)dy = 0 \tag{1}$$

Now, the differential equation of (1) is a complete (exact) differential equation in the form of $P(x, y)dx + Q(x, y)dy = 0$. Therefore:

$$\Delta = Q_X - P_y = 0$$

$$\Rightarrow 2(1 + \alpha)x^\alpha y^\alpha e^{x^2y^2} + 2\left(2xy^2 e^{x^2y^2}\right)x^{1+\alpha}y^\alpha + 3(1 + \alpha)x^\alpha y^{1+\alpha} - 2(1 + \alpha)x^\alpha y^\alpha e^{x^2y^2}$$

$$- 2\left(2yx^2 e^{x^2y^2}\right)x^\alpha y^{1+\alpha} - 2(2 + \alpha)x^\alpha y^{1+\alpha} = 0$$

$$\Rightarrow 4x^{2+\alpha}y^{2+\alpha}e^{x^2y^2} + 3(1 + \alpha)x^\alpha y^{1+\alpha} - 4y^{2+\alpha}x^{2+\alpha}e^{x^2y^2} - 2(2 + \alpha)x^\alpha y^{1+\alpha} = 0$$

$$\Rightarrow (\alpha - 1)x^\alpha y^{1+\alpha} = 0$$

$$\Rightarrow \alpha = 1$$

Choice (4) is the answer.

2.19. Based on the information given in the problem, we have:

$$e^y dx + (xe^y + 2y)dy = 0 \tag{1}$$

The two-variable differential equation of (1) is a complete differential equation in the form of $P(x, y)dx + Q(x, y)dy = 0$ if $\Delta = Q_x - P_y = 0$. To solve such a differential equation, any term of $P(x, y)$ that include y or any term of $Q(x, y)$ that include x need to be removed, and the remaining terms need to be integrated with respect to the related variable. The problem can be solved as follows.

$$\Delta = Q_x - P_y = e^y - e^y = 0$$

Therefore, the differential equation of (1) is a complete differential equation. Now, the terms of $Q(x, y)$ that include x are removed as follows.

$$e^y dx + 2y dy = 0$$

$$\int \Rightarrow xe^y + y^2 + c = 0$$

Choice (1) is the answer.

2.20. Based on the information given in the problem, we have:

$$\left(x^2 + xy\sin 2x + y\sin^2 x\right)dx + x\sin^2 x\, dy = 0 \tag{1}$$

The two-variable differential equation of (1) is a complete differential equation in the form of $P(x, y)dx + Q(x, y)dy = 0$ if $\Delta = Q_x - P_y = 0$. To solve such a differential equation, any term of $P(x, y)$ that include y or any term of $Q(x, y)$ that include x need to be removed, and the remaining terms need to be integrated with respect to the related variable. The problem can be solved as follows.

$$\Delta = Q_x - P_y = \left(\sin^2 x + x\sin 2x\right) - \left(x\sin 2x + \sin^2 x\right) = 0$$

Therefore, the differential equation of (1) is a complete differential equation. Now, the terms of $P(x, y)$ that include y are removed as follows.

$$\Rightarrow x^2 dx + x\sin^2 x\, dy = 0$$

$$\int \Rightarrow \frac{x^3}{3} + xy\sin^2 x = c$$

Choice (1) is the answer.

2.21. A differential equation with the form of (1) is called a Bernoulli differential equation with respect to y based on x.

$$y'(x) + p(x)y(x) = q(x)y^n(x) \tag{1}$$

A Bernoulli differential equation can be solved as follows.

$$\xrightarrow{\times y^{-n}(x)} y^{-n}(x)y'(x) + p(x)y^{1-n}(x) = q(x) \tag{2}$$

Defining a new variable:

$$y^{1-n}(x) \triangleq u(x) \tag{3}$$

$$\xrightarrow{\frac{d}{dx}} (1-n)y^{-n}(x)y'(x) = u'(x) \Rightarrow y^{-n}(x)y'(x) = \frac{1}{1-n}u'(x) \tag{4}$$

Solving (2)–(4):

$$\frac{1}{1-n}u'(x) + p(x)u(x) = q(x) \Rightarrow u'(x) + (1-n)p(x)u(x) = (1-n)q(x) \tag{5}$$

Equation (5) is a linear equation with respect to y based on x that can be solved as follows.

$$u(x) = \frac{1}{\mu(x)}\{\mu(x)(1-n)q(x)dx + c\} \tag{6}$$

where

$$\mu(x) = e^{\int(1-n)p(x)dx} \tag{7}$$

Solving (3) and (6):

$$y(x) = \left(\frac{1}{\mu(x)}\{\mu(x)(1-n)q(x)dx + c\}\right)^{\frac{1}{1-n}} \tag{8}$$

In this problem, the differential equation is a Bernoulli differential equation with respect to y based on x that can be solved as follows.

$$x\frac{dy}{dx} + y = xy^3 \xrightarrow{\times\frac{y^{-3}}{x}} y^{-3}y' + \frac{1}{x}y^{-2} = 1 \tag{9}$$

By defining a new variable, we have:

$$y^{-2} \triangleq u \tag{10}$$

$$\xrightarrow{\frac{d}{dx}} -2y^{-3}y' = u' \Rightarrow y^{-3}y' = \frac{u'}{-2} \tag{11}$$

Solving (9)–(11):

$$\frac{u'}{-2} + \frac{1}{x}u = 1 \xrightarrow{\times(-2)} u' - \frac{2}{x}u = -2 \tag{12}$$

Equation (12) is a linear differential equation that can be solved as follows.

$$\Rightarrow \mu = e^{\int\frac{-2}{x}dx} = e^{-2\ln x} = \frac{1}{x^2}$$

$$u = \frac{1}{\frac{1}{x^2}}\left\{\int\frac{1}{x^2}\times(-2)dx + c\right\}u(x) = x^2\left(\frac{2}{x} + c\right) \Rightarrow u = 2x + cx^2 \tag{13}$$

Solving (10) and (13):

$$y^{-2} = 2x + cx^2 \Rightarrow y^2 = \frac{1}{2x + cx^2}$$

Choice (2) is the answer.

2.22. The problem can be solved by defining a new variable as follows.

$$(x+y)dy + (x-y)dy = 0 \tag{1}$$

$$\frac{y}{x} \triangleq u \Rightarrow y = ux \tag{2}$$

$$\overset{\frac{d}{dx}}{\Rightarrow} dy = x\,du + u\,dx \tag{3}$$

By solving (1), (2), and (3), the problem becomes a separable differential equation.

$$(x+xu)(x\,du + u\,dx) + (x-xu)dx = 0 \Rightarrow (1+u)x\,du + (1+u^2)dx = 0$$

$$\Rightarrow (1+u)x\,du = -(1+u^2)dx \Rightarrow \frac{1+u}{1+u^2}du = -\frac{dx}{x} \Rightarrow \int \left(\frac{1}{1+u^2} + \frac{u}{1+u^2}\right)du = -\int \frac{dx}{x}dx$$

$$\Rightarrow \tan^{-1}u + \frac{1}{2}\ln(1+u^2) = -\ln x + c \Rightarrow 2\tan^{-1}u + \ln(1+u^2) + 2\ln x = c \tag{4}$$

Solving (2) and (4):

$$2\tan^{-1}\left(\frac{y}{x}\right) + \ln\left(1 + \left(\frac{y}{x}\right)^2\right) + \ln x^2 = c \Rightarrow 2\tan^{-1}\left(\frac{y}{x}\right) + \ln\left(x^2 + \left(\frac{y}{x}\right)^2 x^2\right) = c$$

$$\Rightarrow 2\tan^{-1}\left(\frac{y}{x}\right) + \ln(x^2 + y^2) = c$$

Choice (3) is the answer.

2.23. A differential equation with the form of (1) is called a linear differential equation with respect to y based on x.

$$y'(x) + p(x)y(x) = q(x) \tag{1}$$

The solution of a linear differential equation can be determined as follows.

$$y(x) = \frac{1}{\mu(x)}\{\mu(x)q(x)dx + c\} \tag{2}$$

where

$$\mu(x) = e^{\int p(x)dx} \tag{3}$$

Based on the information given in the problem, we have:

$$(\cos x)y' + (\sin x)y = 2x\cos^2 x \tag{4}$$

$$y\left(\frac{\pi}{4}\right) = -\frac{15\sqrt{2}\pi^2}{32} \tag{5}$$

The problem can be solved as follows.

$$(\cos x)y' + (\sin x)y = 2x\cos^2 x \xrightarrow{\times \frac{1}{\cos x}} y' + (\tan x)y = 2x\cos x$$

Now, the problem is in the form of a linear differential equation that can be solved as follows.

$$\Rightarrow \mu(x) = e^{\int \tan x \, dx} = e^{-\ln \cos x} = \frac{1}{\cos x}$$

$$y(x) = \frac{1}{\frac{1}{\cos x}}\left\{\int \frac{1}{\cos x} \times (2x\cos x)\, dx + c\right\} \Rightarrow y(x) = (x^2 + c)\cos x \tag{6}$$

Solving (5) and (6):

$$-\frac{15\sqrt{2}\pi^2}{32} = \left(\frac{\pi^2}{16} + c\right)\frac{\sqrt{2}}{2} \Rightarrow -\frac{15\pi^2}{16} = \frac{\pi^2}{16} + c \Rightarrow c = -\pi^2 \tag{7}$$

Solving (6) and (7):

$$\Rightarrow y(x) = (x^2 - \pi^2)\cos x$$

Therefore:

$$y(0) = (0 - \pi^2) \times 1 \Rightarrow y(0) = -\pi^2$$

Choice (4) is the answer.

2.24. **The first method**: The differential equation is a Bernoulli differential equation with respect to y based on x that can be solved as follows.

$$x\, dy - y\, dx = xy^2 dx \xrightarrow{\times \frac{1}{x\, dx}} \frac{dy}{dx} - \frac{1}{x}y = y^2 \xrightarrow{\times y^{-2}} y^{-2}y' - \frac{1}{x}y^{-1} = 1 \tag{1}$$

By defining a new variable, we have:

$$y^{-1} \triangleq u \tag{2}$$

$$\xRightarrow{\frac{d}{dx}} y^{-2}y' = -u' \tag{3}$$

Solving (1)–(3):

$$-u' - \frac{1}{x}u = 1 \Rightarrow u' + \frac{1}{x}u = -1$$

Now, the problem is in the form of a linear differential equation that can be solved as follows.

$$\Rightarrow \mu(x) = e^{\int \frac{1}{x}dx} = e^{\ln x} = x$$

$$u(x) = \frac{1}{x}\left\{ \int x \times (-1)dx + c \right\} \Rightarrow u(x) = \frac{1}{x}\left(-\frac{x^2}{2} + c \right) \tag{4}$$

Solving (2) and (4):

$$\frac{1}{y} = \frac{1}{x}\left(-\frac{x^2}{2} + c \right) \Rightarrow \frac{x}{y} = -\frac{x^2}{2} + c \Rightarrow \frac{x^2}{2} + \frac{x}{y} = c$$

The second method: The differential equation can be rearranged in the form of a complete (exact) differential equation as follows.

$$x\,dy - y\,dx = xy^2 dx \Rightarrow \frac{x\,dy - y\,dx}{y^2} = x\,dx$$

$$\Rightarrow -d\left(\frac{x}{y} \right) = d\left(\frac{x^2}{2} \right) \Rightarrow d\left(\frac{x^2}{2} + \frac{x}{y} \right) = 0 \xrightarrow{\int} \frac{x^2}{2} + \frac{x}{2} = c$$

Choice (3) is the answer.

2.25. The differential equation is a Bernoulli differential equation with respect to y based on x that can be solved as follows.

$$y' + y = \frac{x}{y} \Rightarrow y' + y = xy^{-1} \xrightarrow{xy} yy' + y^2 = x \tag{1}$$

By defining a new variable, we have:

$$y^2 \triangleq u \tag{2}$$

$$\xrightarrow{\frac{d}{dx}} 2yy' = u' \tag{3}$$

Solving (1)–(3):

$$\frac{u'}{2} + u = x \Rightarrow u' + 2u = 2x$$

Now, the problem is in the form of a linear differential equation that can be solved as follows.

$$\mu(x) = e^{\int 2dx} = e^{2x}$$

$$u(x) = \frac{1}{e^{2x}}\left\{ \int e^{2x} \times (2x)dx + c \right\} = e^{-2x}\left(\left(x - \frac{1}{2} \right)e^{2x} + c \right) = x - \frac{1}{2} + ce^{-2x} \tag{4}$$

Solving (2) and (4):

$$y^2 = x - \frac{1}{2} + ce^{-2x}$$

Choice (4) is the answer.

2.26. The differential equation is a Bernoulli differential equation with respect to y based on x that can be solved as follows.

$$xy' + y = x^4 y^3 \Rightarrow y' + \left(\frac{1}{x}\right)y = x^3 y^3 \xrightarrow{\times y^{-3}} y^{-3}y' + \left(\frac{1}{x}\right)y^{-2} = x^3$$

By defining a new variable, we have:

$$y^{-2} \triangleq u \tag{2}$$

$$\xrightarrow{\frac{d}{dx}} -2y^{-3}y' = u' \tag{3}$$

Solving (1)–(3):

$$-\frac{u'}{2} + \frac{1}{x}u = x^3 \Rightarrow u' - \left(\frac{2}{x}\right)u = -2x^3$$

Now, the problem is in the form of a linear differential equation that can be solved as follows.

$$\mu(x) = e^{\int -\frac{2}{x}dx} = e^{-2\ln x} = \frac{1}{x^2}$$

$$u(x) = \frac{1}{\frac{1}{x^2}}\left\{ \int \frac{1}{x^2} \times (-2x^3)dx + c \right\} = x^2\left(-x^2 + c\right) \tag{4}$$

Solving (2) and (4):

$$y^{-2} = x^2\left(-x^2 + c\right)$$

Choice (2) is the answer.

2.27. Based on the information given in the problem, we have:

$$y' = xy^2 - y \tag{1}$$

$$y(x = 0) = 1 \tag{2}$$

The differential equation is a Bernoulli differential equation with respect to y based on x that can be solved as follows.

$$y' = xy^2 - y \Rightarrow y' + y = xy^2 \xrightarrow{\times y^{-2}} y^{-2}y' + y^{-1} = x \tag{3}$$

By defining a new variable, we have:

$$y^{-1} \triangleq u \tag{4}$$

$$\xrightarrow{\frac{d}{dx}} -y^{-2}y' = u' \tag{5}$$

Solving (3)–(5):

$$-u' + u = x \Rightarrow u' - u = -x$$

Now, the problem is in the form of a linear differential equation that can be solved as follows.

$$\mu(x) = e^{\int -dx} = e^{-x}$$

$$u(x) = \frac{1}{e^{-x}}\left\{\int e^{-x} \times (-x)dx + c\right\} = e^{x}(xe^{-x} + e^{-x} + c) = x + 1 + ce^{x} \tag{6}$$

Solving (4) and (6):

$$y^{-1} = x + 1 + ce^{x} \Rightarrow y = \frac{1}{x + 1 + ce^{x}} \tag{7}$$

Solving (2) and (7):

$$1 = \frac{1}{1+c} \Rightarrow c = 0 \tag{8}$$

Solving (7) and (8):

$$y = \frac{1}{x+1}$$

Choice (1) is the answer.

2.28. The differential equation is a Bernoulli differential equation with respect to y based on x that can be solved as follows.

$$y' - \frac{1}{x}y = -\frac{1}{x}y^{2} \xrightarrow{\times y^{-2}} y^{-2}y' - \frac{1}{x}y^{-1} = -\frac{1}{x} \tag{1}$$

By defining a new variable, we have:

$$y^{-1} \triangleq u \tag{2}$$

$$\xrightarrow{\frac{d}{dx}} -y^{-2}y' = u' \tag{3}$$

Solving (1)–(3):

$$-u' - \frac{1}{x}u = -\frac{1}{x} \Rightarrow u' + \frac{1}{x}u = \frac{1}{x}$$

Now, the problem is in the form of a linear differential equation that can be solved as follows.

$$\mu(x) = e^{\int \frac{1}{x}dx} = e^{\ln x} = x$$

$$u(x) = \frac{1}{x}\left\{\int x \times \frac{1}{x}dx + c\right\} = \frac{1}{x}(x + c) \tag{4}$$

Solving (2) and (4):

$$\frac{1}{y} = \frac{x+c}{x} \Rightarrow y = \frac{x}{x+c}$$

Choice (3) is the answer.

2.29. The problem can be solved as follows.

$$y' = \sin(x - y) \qquad (1)$$

By defining a new variable, we have:

$$x - y \triangleq u \qquad (2)$$

$$\Rightarrow y' = 1 - u' \qquad (3)$$

Solving (1)–(3):

$$1 - u' = \sin u \Rightarrow \frac{du}{dx} = 1 - \sin u \Rightarrow \frac{du}{1 - \sin u} = dx$$

$$\Rightarrow \int \frac{du}{1 - \sin u} = \int dx \qquad (4)$$

As we know:

$$\int \frac{du}{1 - \sin u} = \int \frac{1 + \sin u}{1 - \sin^2 u} du = \int \frac{1 + \sin u}{\cos^2 u} du = \int \frac{du}{\cos^2 u} + \int \frac{\sin u}{\cos^2 u} du = \tan u + \frac{1}{\cos u} = \tan u + \sec u \qquad (5)$$

Solving (4) and (5):

$$\tan u + \sec u = x + c \qquad (6)$$

Solving (2) and (6):

$$\tan(x - y) + \sec(x - y) = x + c$$

Choice (2) is the answer.

2.30. The differential equation is a Bernoulli differential equation with respect to x based on y that can be solved as follows.

$$y\, dx + x(x^2 y - 1) dy = 0 \Rightarrow \frac{dx}{dy} = \frac{x - x^3 y}{y} \Rightarrow \frac{dx}{dy} - \frac{1}{y} x = -x^3 \xrightarrow{\times x^{-3}} x^{-3} \frac{dx}{dy} - \frac{1}{y} x^{-2} = -1 \qquad (1)$$

By defining a new variable, we have:

$$x^{-2} = u(y) \qquad (2)$$

$$\xrightarrow{\frac{d}{dy}} x^{-3} \frac{dx}{dy} = -\frac{1}{2} \frac{du}{dy} \qquad (3)$$

Solving (1)–(3):

$$-\frac{1}{2} \frac{du}{dy} - \frac{1}{y} u = -1 \Rightarrow \frac{du}{dy} + \frac{2}{y} u = 2$$

Now, the problem is in the form of a linear differential equation with respect to x based on y that can be solved as follows.

$$\mu(y) = e^{\int \frac{2}{y}dy} = e^{2\ln y} = y^2$$

$$u(y) = \frac{1}{y^2}\left\{\int y^2 \times (2)dy + C\right\} = \frac{1}{y^2}\left(\frac{2y^3}{3} + C\right) \tag{4}$$

Solving (2) and (4):

$$\frac{1}{x^2} = \frac{1}{y^2}\left(\frac{2}{3}y^3 + C\right) \Rightarrow \frac{y^2}{x^2} = \frac{2}{3}y^3 + C \Rightarrow \frac{y^2}{2x^2} = \frac{y^3}{3} + \frac{C}{2} \Rightarrow -\frac{y^2}{2x^2} + \frac{y^3}{3} = -\frac{C}{2}$$

$$\Rightarrow -\frac{y^2}{2x^2} + \frac{y^3}{3} = c$$

Choice (2) is the answer.

Problems: Second-Order Differential Equations

3

Abstract

In this chapter, different types of second-order differential equations with constant coefficients, including homogeneous and non-homogeneous differential equations, are studied. In this regard, the second-order differential equations are solved by using inverse operator method, undetermined coefficient method, and Lagrange method. In this chapter, the problems are categorized in different levels based on their difficulty levels (easy, normal, and hard) and calculation amounts (small, normal, and large). Additionally, the problems are ordered from the easiest problem with the smallest computations to the most difficult problems with the largest calculations.

3.1. Determine the general solution of the following homogeneous second-order differential equation.

$$y'' - 2y' + y = 0$$

Difficulty level   ● Easy   ○ Normal   ○ Hard
Calculation amount   ● Small   ○ Normal   ○ Large

1) $cx^2 e^x$
2) cxe^x
3) $c_1 xe^x + c_2 x$
4) $(c_1 + c_2 x)e^x$

3.2. Solve the homogeneous second-order differential equation below.

$$y'' - 2y' = 0$$

Difficulty level   ● Easy   ○ Normal   ○ Hard
Calculation amount   ● Small   ○ Normal   ○ Large

1) $c_1 e^{-x} + c_2 e^x$
2) $c_1 e^{-x} + c_2 x e^{x^2}$
3) $c_1 xe^x + c_2 x^2 e^x$
4) $c_1 + c_2 e^{2x}$

3.3. Determine the homogeneous solution of the second-order differential equation below.

$$y'' + 3y' + 2y = 2x + 1$$

Difficulty level   ● Easy   ○ Normal   ○ Hard
Calculation amount   ● Small   ○ Normal   ○ Large

1) $c_1 e^{-x} + c_2 e^{-2x}$
2) $c_1 e^{-2x} + c_2 e^{x}$
3) $c_1 e^{-x} + c_2 x e^{-2x}$
4) $c_1 x e^{-x} + c_2 x^2 e^{-2x}$

3.4. Solve the homogeneous second-order differential equation below.

$$y''' - y' = 0$$

Difficulty level   ● Easy   ○ Normal   ○ Hard
Calculation amount   ● Small   ○ Normal   ○ Large

1) $c_1 e^{x} + c_2 e^{-x} + c_3$
2) $(c_1 + c_2 x) e^{x}$
3) $c_1 \sin x + c_2 \cos x + c_3$
4) $x(c_1 e^{x} + c_2 e^{-x})$

3.5. Solve the homogeneous second-order differential equation below.

$$y''' - y = 0$$

Difficulty level   ○ Easy   ● Normal   ○ Hard
Calculation amount   ○ Small   ● Normal   ○ Large

1) $c_1 e^{x} + e^{\frac{-x}{2}} \left(c_2 \cos \left(\frac{\sqrt{3}}{2} x \right) + c_3 \sin \left(\frac{\sqrt{3}}{2} x \right) \right)$
2) $c_1 e^{x} + e^{\frac{x}{2}} \left(c_2 \cos \left(\sqrt{2} x \right) + c_3 \sin \left(\sqrt{2} x \right) \right)$
3) $c_1 e^{-x} + e^{\frac{x}{2}} \left(c_2 \cos \left(\sqrt{3} x \right) + c_3 \sin \left(\sqrt{3} x \right) \right)$
4) $c_1 e^{-x} + e^{\frac{x}{2}} \left(c_2 \cos \left(\frac{\sqrt{2}}{2} x \right) + c_3 \sin \left(\frac{\sqrt{2}}{2} x \right) \right)$

3.6. Solve the homogeneous second-order differential equation below with the given primary conditions.

$$y'' + y' - 2y = 0, \quad y(x = 0) = 4, \quad y'(x = 0) = 1$$

Difficulty level   ○ Easy   ● Normal   ○ Hard
Calculation amount   ○ Small   ● Normal   ○ Large

1) $3e^{x} - e^{-2x}$
2) $\frac{3}{2} e^{x}$
3) $3e^{x} - x$
4) $3e^{x} + e^{-2x}$

3.7. Solve the homogeneous second-order differential equation below with the given primary conditions.

$$y'' - 4y' + 4y = 0, \quad y(x = 0) = 3, \quad y'(x = 0) = 4$$

Difficulty level   ○ Easy   ● Normal   ○ Hard
Calculation amount   ○ Small   ● Normal   ○ Large

1) $(3 - 2x) e^{2x}$
2) $(3 + 2x) e^{2x}$
3) $(3 - 2x) e^{x}$
4) $(-3 + 2x) e^{2x}$

3.8. Which one of the following choices is one of the particular solutions of the non-homogeneous second-order differential equation below?

$$y'' + 2y' - 8y = e^{3x}$$

Difficulty level   ○ Easy   ● Normal   ○ Hard
Calculation amount   ○ Small   ● Normal   ○ Large

1) $\frac{1}{9}e^{3x}$

2) $\frac{1}{7}xe^{3x}$

3) $\frac{1}{7}e^{3x}$

4) $\frac{1}{9}xe^{3x}$

3.9. Calculate the general solution of the non-homogeneous second-order differential equation below.

$$y'' - 4y' + 3y = 10e^{-2x}$$

Difficulty level   ○ Easy   ● Normal   ○ Hard
Calculation amount   ○ Small   ● Normal   ○ Large

1) $c_1 e^x + c_2 e^{3x} + \frac{2}{3}e^{-2x}$

2) $c_1 e^x + c_2 e^{3x} + c_3 e^{-2x}$

3) $c_1 e^x + c_2 e^{3x}$

4) ce^{-2x}

3.10. Calculate the general solution of the non-homogeneous second-order differential equation below.

$$y'' - 4y' + 3y = e^{3x}$$

Difficulty level   ○ Easy   ● Normal   ○ Hard
Calculation amount   ○ Small   ● Normal   ○ Large

1) $y = c_1 e^x + c_2 e^{2x} + \frac{x}{2}e^{3x}$

2) $y = c_1 e^x + c_2 e^{2x} + xe^{4x}$

3) $y = c_1 e^{2x} + c_2 e^{4x} + \frac{x^2}{2}e^{3x}$

4) $y = c_1 e^x + c_2 e^{3x} + \frac{x}{2}e^{3x}$

3.11. Calculate the particular solution of the non-homogeneous second-order differential equation below.

$$y'' + 4y = 3\sin(4x)$$

Difficulty level   ○ Easy   ● Normal   ○ Hard
Calculation amount   ○ Small   ● Normal   ○ Large

1) $-\frac{1}{4}\sin(4x)$

2) $\frac{1}{4}\sin(4x)$

3) $-\frac{3}{20}\sin(4x)$

4) $\frac{3}{20}\sin(4x)$

3.12. Calculate the particular solution of the non-homogeneous second-order differential equation below.

$$y'' + 3y' + 3y = \sin(2x)$$

Difficulty level   ○ Easy   ● Normal   ○ Hard
Calculation amount   ○ Small   ● Normal   ○ Large

1) $-\frac{1}{37}(6\cos{(2x)} + \sin{(2x)})$

2) $\frac{1}{37}(6\cos{(2x)} + \sin{(2x)})$

3) $-\frac{1}{37}\sin{(2x)}$

4) $\frac{1}{37}\sin{(2x)}$

3.13. Determine the general solution of the non-homogeneous second-order differential equation below.

$$\frac{d^4 y}{dx^4} - y = x\sin x$$

Difficulty level   ○ Easy   ● Normal   ○ Hard
Calculation amount   ○ Small   ● Normal   ○ Large

1) $(c_1 + c_2 x)e^x + (c_3 + c_4 x)e^{-x} + (c_5 + c_6 x)\sin x + (c_7 + c_8 x)\cos x$

2) $c_1 e^x + c_2 e^{-x} + c_3 \sin x + c_4 \cos x + x(c_5 + c_6 x)\sin x + x(c_7 + c_8 x)\cos x$

3) $c_1 e^x + c_2 e^{-x} + c_3 \sin x + c_4 \cos x + (c_5 + c_6 x)\sin x + (c_7 + c_8 x)\cos x$

4) $(c_1 + c_2 x)e^x + c_3 \sin x + c_4 \cos x + x(c_5 + c_6 x)\sin x + x(c_7 + c_8 x)\cos x$

3.14. Solve the following homogeneous second-order differential equation.

$$y'' - 5y' + 7y = 0$$

Difficulty level   ○ Easy   ● Normal   ○ Hard
Calculation amount   ○ Small   ● Normal   ○ Large

1) $e^{\frac{5}{2}x}\left(c_1 \sin\left(\frac{\sqrt{5}}{2}x\right) + c_2 \cos\left(\frac{\sqrt{5}}{2}x\right)\right)$

2) $e^{\frac{5}{2}x}\left(c_1 \sin\left(\frac{\sqrt{3}}{2}x\right) + c_2 \cos\left(\frac{\sqrt{3}}{2}x\right)\right)$

3) $e^{\frac{5}{2}x}\left(c_1 \sin\left(\frac{\sqrt{3}}{2}x\right) + c_2 x\cos\left(\frac{\sqrt{3}}{2}x\right)\right)$

4) $e^{\frac{5}{2}x}\left(c_1 \sin\left(\frac{\sqrt{5}}{2}x\right) + c_2 x^2 \cos\left(\frac{\sqrt{5}}{2}x\right)\right)$

3.15. Calculate the particular solution of the non-homogeneous second-order differential equation below.

$$y'' + y = x^2$$

Difficulty level   ○ Easy   ○ Normal   ● Hard
Calculation amount   ○ Small   ● Normal   ○ Large

1) x^2

2) $x^2 + 2$

3) $x^2 - 1$

4) $x^2 - 2$

3.16. Calculate the general solution of the non-homogeneous second-order differential equation below.

$$y'' - 2y' + 2y = 2x - 2$$

Difficulty level   ○ Easy   ○ Normal   ● Hard
Calculation amount   ○ Small   ● Normal   ○ Large

1) $y = Ae^{-x}\sin(x + \alpha) + x$

2) $y = Ae^x \sin(x + \alpha) - x$

3) $y = Ae^x \sin(x + \alpha) + x$

4) $y = Ae^{-x}\sin(x + \alpha) - x$

3.17. What is the particular solution of the non-homogeneous second-order differential equation below?

$$y'' - y = 8xe^x$$

Difficulty level   ○ Easy   ○ Normal   ● Hard
Calculation amount   ○ Small   ● Normal   ○ Large

1) $y = \cos x$
2) $y = \frac{x}{2} \sin x$
3) $y = \frac{x^2}{2} e^{2x}$
4) $y = (1 - 2x + 2x^2)e^x$

3.18. Which one of the following choices is one of the particular solutions of the non-homogeneous second-order differential equation below?

$$y'' + 2y' + 3y = x \cos x$$

Difficulty level   ○ Easy   ○ Normal   ● Hard
Calculation amount   ○ Small   ● Normal   ○ Large

1) $\frac{1}{4}(x - 1)(\sin x + \cos x)$
2) $\frac{1}{4}(x - 1)(\sin x - \cos x)$
3) $\frac{1}{4}(x - 1)(-\sin x + \cos x)$
4) $\frac{1}{4}(x - 1)(-\sin x - \cos x)$

3.19. Determine the general solution of the non-homogeneous second-order differential equation below.

$$\left(D^2 - 1\right)y = 8xe^x$$

Difficulty level   ○ Easy   ○ Normal   ● Hard
Calculation amount   ○ Small   ● Normal   ○ Large

1) $y = c_1 e^x + c_2 e^{-x} + \cos x$
2) $y = c_1 \cos x + c_2 \sin x + \frac{x}{2} \sin x$
3) $y = e^{2x}\left(c_1 + c_2 x + \frac{x^2}{2}\right)$
4) $y = c_1 e^{-x} + e^x(c_2 - 2x + 2x^2)$

3.20. Calculate the general solution of the non-homogeneous second-order differential equation below.

$$y'' - 4y' + 4y = \frac{e^{2x}}{x}$$

Difficulty level   ○ Easy   ○ Normal   ● Hard
Calculation amount   ○ Small   ○ Normal   ● Large

1) $(c_1 + x \ln x)e^{2x}$
2) $(c_1 + c_2 x + x - \ln x)e^{2x}$
3) $\left(c_1 + c_2 x + \frac{1}{x} - \ln x\right)e^{2x}$
4) $(c_1 + c_2 x + x \ln x - x)e^{2x}$

3.21. Calculate the value of $y(x = -1)$ in the non-homogeneous second-order differential equation below with the given primary conditions.

$$y'' + 6y' + 9y = e^{-3x} \cosh x, \quad y(x = 0) = 1, \quad y(x = 1) = 0$$

Difficulty level   ○ Easy   ○ Normal   ● Hard
Calculation amount   ○ Small   ○ Normal   ● Large

1) $2e^{-3} \cosh(1)$

2) $e^9 \cosh(1)$

3) $e^{-9} \cosh(1)$

4) $2e^3 \cosh(1)$

3.22. Calculate the value of $y\left(x = \frac{\pi}{4}\right)$ in the non-homogeneous second-order differential equation below with the given primary conditions.

$$y'' + 4y' = \sin 2x, \quad y(x = 0) = y'(x = 0) = 1$$

Difficulty level   ○ Easy   ○ Normal   ● Hard
Calculation amount   ○ Small   ○ Normal   ● Large

1) $\frac{8}{5}$

2) $\frac{5}{8}$

3) 8

4) 5

3.23. Calculate the value of $y(x = \pi)$ in the homogeneous second-order differential equation below with the given primary conditions.

$$y^{(4)} + 5y'' + 4y = 0, \quad y(x = 0) = 1, y'(x = 0) = -1, y''(x = 0) = 1, y'''(x = 0) = 0$$

Difficulty level   ○ Easy   ○ Normal   ● Hard
Calculation amount   ○ Small   ○ Normal   ● Large

1) $\frac{7}{3}$

2) $-\frac{7}{3}$

3) $\frac{3}{5}$

4) $-\frac{5}{3}$

Solutions of Problems: Second-Order Differential Equations

4

Abstract

In this chapter, the problems of the third chapter are fully solved, in detail, step-by-step, and with different methods.

4.1. First, we need to determine the roots of the characteristic equation of differential equation as follows.

$$y'' - 2y' + y = 0 \Rightarrow \lambda^2 - 2\lambda + 1 = 0 \Rightarrow (\lambda - 1)^2 = 0 \Rightarrow \lambda = 1, 1$$

Therefore, the general solution of the homogeneous differential equation is as follows.

$$y_h = (c_1 + c_2 x)e^x$$

Choice (4) is the answer.

4.2. First, we need to determine the roots of the characteristic equation of differential equation as follows.

$$y'' - 2y' = 0 \Rightarrow \lambda^2 - 2\lambda = 0 \Rightarrow \lambda(\lambda - 2) = 0 \Rightarrow \lambda = 0, 2$$

Therefore, the general solution of the homogeneous differential equation is as follows.

$$y = c_1 + c_2 e^{2x}$$

Choice (4) is the answer.

4.3. First, we need to determine the roots of the characteristic equation of differential equation as follows.

$$y'' + 3y' + 2y = 2x + 1 \Rightarrow \lambda^2 + 3\lambda + 2 = 0 \Rightarrow (\lambda + 1)(\lambda + 2) = 0 \Rightarrow \lambda = -1, -2$$

Therefore, the general solution of the homogeneous differential equation is as follows.

$$y_h = c_1 e^{-x} + c_2 e^{-2x}$$

Choice (1) is the answer.

4.4. First, we need to determine the roots of the characteristic equation of differential equation as follows.

$$y''' - y' = 0 \Rightarrow \lambda^3 - \lambda = 0 \Rightarrow \lambda(\lambda^2 - 1) = 0 \Rightarrow \lambda = 0, -1, 1$$

Therefore, the general solution of the homogeneous differential equation is as follows.

$$y = c_1 e^x + c_2 e^{-x} + c_3$$

Choice (1) is the answer.

4.5. First, we need to determine the roots of the characteristic equation of differential equation as follows.

$$y''' - y = 0 \Rightarrow \lambda^3 - 1 = 0 \Rightarrow (\lambda - 1)(\lambda^2 + \lambda + 1) = 0$$

$$\Rightarrow \begin{cases} \lambda - 1 = 0 \Rightarrow \lambda = 1 \\ \lambda^2 + \lambda + 1 = 0 \Rightarrow \lambda = \dfrac{-1 \pm \sqrt{1-4}}{2} = \dfrac{-1}{2} \pm i\dfrac{\sqrt{3}}{2} \end{cases}$$

Therefore, the general solution of the homogeneous differential equation is as follows.

$$\Rightarrow y = c_1 e^x + e^{\frac{-x}{2}}\left(c_2 \cos\left(\frac{\sqrt{3}}{2}x\right) + c_3 \sin\left(\frac{\sqrt{3}}{2}x\right)\right)$$

Choice (1) is the answer.

4.6. Based on the information given in the problem, we have:

$$y(x = 0) = 4 \tag{1}$$

$$y'(x = 0) = 1 \tag{2}$$

$$y'' + y' - 2y = 0 \tag{3}$$

First, we need to determine the roots of the characteristic equation of differential equation as follows.

$$\Rightarrow \lambda^2 + \lambda - 2 = 0 \Rightarrow (\lambda + 2)(\lambda - 1) = 0 \Rightarrow \begin{cases} \lambda = 1 \\ \lambda = -2 \end{cases} \tag{4}$$

Therefore, the general solution of the homogeneous differential equation is as follows.

$$y = c_1 e^x + c_2 e^{-2x} \tag{5}$$

$$y' = c_1 e^x - 2c_2 e^{-2x} \tag{6}$$

Solving (1) and (5):

$$c_1 + c_2 = 4 \tag{7}$$

Solving (2) and (6):

$$c_1 - 2c_2 = 1 \tag{8}$$

Solving (7) and (8):

$$c_1 = 3, \quad c_2 = 1 \tag{9}$$

Solving (5) and (9):

$$y = 3e^x + e^{-2x}$$

Choice (4) is the answer.

4.7. Based on the information given in the problem, we have:

$$y(x = 0) = 3 \tag{1}$$

$$y'(x = 0) = 4 \tag{2}$$

$$y'' - 4y' + 4y = 0 \tag{3}$$

First, we need to determine the roots of the characteristic equation of differential equation as follows.

$$\lambda^2 - 4\lambda + 4 = 0 \Rightarrow (\lambda - 2)^2 = 0 \Rightarrow \lambda = 2, 2 \tag{4}$$

Therefore, the general solution of the homogeneous differential equation is as follows.

$$y = c_1 e^{2x} + c_2 x e^{2x} \tag{5}$$

$$y' = 2c_1 e^{2x} + c_2 e^{2x} + 2c_2 x e^{2x} \tag{6}$$

Solving (1) and (5):

$$c_1 = 3 \tag{7}$$

Solving (2) and (6):

$$2c_1 + c_2 = 4 \xrightarrow{\text{Using (7)}} c_2 = -2 \tag{8}$$

Solving (5), (7), and (8):

$$y = (3 - 2x)e^{2x}$$

Choice (1) is the answer.

4.8. Suppose that the non-homogeneous second-order differential equation is in the form below:

$$a_n y^{(n)}(x) + a_{n-1} y^{(n-1)}(x) + \ldots + a_1 y'(x) + a_0 y(x) = r(x)$$

$$\left(a_n D^n + a_{n-1} D^{n-1} + \ldots + a_1 D + a_0 \right) y(x) = r(x)$$

$$F(D)y(x) = r(x)$$

The particular solution of the non-homogeneous second-order differential equation can be solved using inverse operator method. If $r(x) = Ae^{\alpha x}$, we can use the following rule.

$$y_p = \frac{1}{F(D)} e^{\alpha x} = \begin{cases} \dfrac{1}{F(\alpha)} e^{\alpha x} & F(\alpha) \neq 0 \\[2mm] \dfrac{x}{F'(\alpha)} e^{\alpha x} & F(\alpha) = 0, F'(\alpha) \neq 0 \\[2mm] \dfrac{x^2}{F''(\alpha)} e^{\alpha x} & F(\alpha) = F'(\alpha) = 0, F''(\alpha) \neq 0 \\[2mm] \vdots \\[2mm] \dfrac{x^k}{F^{(k)}(\alpha)} e^{\alpha x} & F(\alpha) = F'(\alpha) = \cdots = F^{(k-1)}(\alpha) = 0, F^{(k)}(\alpha) \neq 0 \end{cases}$$

Therefore, for $y'' + 2y' - 8y = e^{3x}$, we can write:

$$y_p = \frac{1}{D^2 + 2D - 8} e^{3x} \Rightarrow y_p = \frac{1}{7} e^{3x}$$

Choice (3) is the answer.

4.9. First, we need to determine the roots of the characteristic equation of differential equation as follows.

$$y'' - 4y' + 3y = 10e^{-2x} \Rightarrow \lambda^2 - 4\lambda + 3 = 0 \Rightarrow (\lambda - 1)(\lambda - 3) = 0 \Rightarrow \lambda = 1, 3$$

Therefore, the general solution of the homogeneous differential equation is as follows.

$$y_h = c_1 e^x + c_2 e^{3x}$$

The particular solution of the non-homogeneous second-order differential equation can be solved using inverse operator method. If $r(x) = A e^{\alpha x}$, we can use the following rule.

$$y_p = \frac{1}{F(D)} e^{\alpha x} = \begin{cases} \dfrac{1}{F(\alpha)} e^{\alpha x} & F(\alpha) \neq 0 \\[2mm] \dfrac{x}{F'(\alpha)} e^{\alpha x} & F(\alpha) = 0, F'(\alpha) \neq 0 \\[2mm] \dfrac{x^2}{F''(\alpha)} e^{\alpha x} & F(\alpha) = F'(\alpha) = 0, F''(\alpha) \neq 0 \\[2mm] \vdots \\[2mm] \dfrac{x^k}{F^{(k)}(\alpha)} e^{\alpha x} & F(\alpha) = F'(\alpha) = \cdots = F^{(k-1)}(\alpha) = 0, F^{(k)}(\alpha) \neq 0 \end{cases}$$

Therefore, for $y'' - 4y' + 3y = 10e^{-2x}$, we can write:

$$y_p = \frac{1}{D^2 - 4D + 3} 10e^{-2x} = \frac{2}{3} e^{-2x}$$

Finally, the general solution of the non-homogeneous second-order differential equation is:

$$y = y_h + y_p = c_1 e^x + c_2 e^{3x} + \frac{2}{3} e^{-2x}$$

Choice (1) is the answer.

4.10. First, we need to determine the roots of the characteristic equation of differential equation as follows.

$$y'' - 4y' + 3y = e^{3x} \Rightarrow \lambda^2 - 4\lambda + 3 = 0 \Rightarrow (\lambda - 1)(\lambda - 3) = 0 \Rightarrow \lambda = 1, 3$$

Therefore, the general solution of the homogeneous differential equation is as follows.

$$y_h = c_1 e^x + c_2 e^{3x}$$

The particular solution of the non-homogeneous second-order differential equation can be solved using inverse operator method. If $r(x) = Ae^{\alpha x}$, we can use the following rule.

$$y_p = \frac{1}{F(D)} e^{\alpha x} = \begin{cases} \dfrac{1}{F(\alpha)} e^{\alpha x} & F(\alpha) \neq 0 \\[2mm] \dfrac{x}{F'(\alpha)} e^{\alpha x} & F(\alpha) = 0, F'(\alpha) \neq 0 \\[2mm] \dfrac{x^2}{F''(\alpha)} e^{\alpha x} & F(\alpha) = F'(\alpha) = 0, F''(\alpha) \neq 0 \\[2mm] \vdots \\[2mm] \dfrac{x^k}{F^{(k)}(\alpha)} e^{\alpha x} & F(\alpha) = F'(\alpha) = \cdots = F^{(k-1)}(\alpha) = 0, F^{(k)}(\alpha) \neq 0 \end{cases}$$

Therefore, for $y'' - 4y' + 3y = e^{3x}$, we have:

$$y_p = \frac{1}{D^2 - 4D + 3} e^{3x} = \frac{x}{2} e^{3x}$$

Finally, the general solution of the non-homogeneous second-order differential equation is:

$$y = y_h + y_p = c_1 e^x + c_2 e^{3x} + \frac{x}{2} e^{3x}$$

Choice (4) is the answer.

4.11. The particular solution of the non-homogeneous second-order differential equation can be solved using inverse operator method. If $r(x) = \cos(\beta x + \gamma)$ or $r(x) = \sin(\beta x + \gamma)$, we can use the following rule.

$$\begin{cases} \dfrac{1}{F(D^2)} \sin(\beta x + \gamma) = \dfrac{1}{F(-\beta^2)} \sin(\beta x + \gamma) \\[3mm] \dfrac{1}{F(D^2)} \cos(\beta x + \gamma) = \dfrac{1}{F(-\beta^2)} \cos(\beta x + \gamma) \end{cases}, \quad F(-\beta^2) \neq 0,$$

Therefore, for $y'' + 4y = 3\sin(4x)$, we have:

$$y_p = \frac{1}{D^2 + 4}(3\sin 4x) = \frac{1}{-16 + 4}(3\sin(4x)) = \frac{-1}{4}\sin(4x)$$

Choice (1) is the answer.

4.12. The particular solution of the non-homogeneous second-order differential equation can be solved using inverse operator method. If $r(x) = \cos(\beta x + y)$ or $r(x) = \sin(\beta x + y)$, we can use the following rule.

$$\begin{cases} \dfrac{1}{F(D^2)} \sin(\beta x + \gamma) = \dfrac{1}{F(-\beta^2)} \sin(\beta x + \gamma) \\ \dfrac{1}{F(D^2)} \cos(\beta x + \gamma) = \dfrac{1}{F(-\beta^2)} \cos(\beta x + \gamma) \end{cases}, F(-\beta^2) \neq 0$$

Therefore, for $y'' + 3y' + 3y = \sin(2x)$, we have:

$$y_p = \frac{1}{D^2 + 3D + 3} \sin(2x) = \frac{1}{-4 + 3D + 3} \sin(2x) = \frac{1}{3D - 1} \sin(2x)$$

$$\xrightarrow{\times \frac{3D+1}{3D+1}} y_p = \frac{3D + 1}{9D^2 - 1} \sin(2x) = \frac{3D + 1}{9(-4) - 1} \sin(2x) = \frac{-1}{37}(3D + 1) \sin(2x)$$

$$y_p = \frac{-1}{37}(6\cos(2x) + \sin(2x))$$

Choice (1) is the answer.

4.13. First, we need to determine the roots of the characteristic equation of differential equation as follows.

$$\frac{d^4 y}{dx^4} - y = x\sin x \Rightarrow \lambda^4 - 1 = 0 \Rightarrow (\lambda^2 - 1)(\lambda^2 + 1) = 0 \Rightarrow \begin{cases} \lambda^2 - 1 = 0 \Rightarrow \lambda = \pm 1 \\ \lambda^2 + 1 = 0 \Rightarrow \lambda = \pm i \end{cases}$$

Therefore, the general solution of the homogeneous differential equation is as follows.

$$y_h = c_1 e^x + c_2 e^{-x} + c_3 \cos x + c_4 \sin x$$

The particular solution of the non-homogeneous second-order differential equation can be solved using undetermined coefficient method. Herein, since $r(x) = x \sin x$, a general guess is made for y_p as follows.

$$y_p = (c_5 x + c_6) \cos x + (c_7 x + c_8) \sin x$$

Since there is a common term in y_p and y_h ($\sin x$ and $\cos x$), y_p needs to be multiplied by x^m, where m is the minimum natural number that removes the similarity between the terms of y_p and y_h. Herein, $m = 1$ satisfies the criterion. Therefore:

$$y_p = x(c_5 x + c_6) \cos x + x(c_7 x + c_8) \sin x$$

Finally, the general solution of the non-homogeneous second-order differential equation is:

$$y = y_h + y_p = c_1 e^x + c_2 e^{-x} + c_3 \cos x + c_4 \sin x + x(c_5 x + c_6) \cos x + x(c_7 x + c_8) \sin x$$

Choice (2) is the answer.

4.14. First, we need to determine the roots of the characteristic equation of differential equation as follows.

$$y'' - 5y' + 7y = 0 \Rightarrow \lambda^2 - 5\lambda + 7 = 0 \Rightarrow \lambda = \frac{5 \pm \sqrt{25 - 28}}{2} = \frac{5}{2} \pm i\frac{\sqrt{3}}{2}$$

Therefore, the general solution of the homogeneous differential equation is as follows.

$$y = e^{\frac{5}{2}x}\left(c_1 \cos\left(\frac{\sqrt{3}}{2}x\right) + c_2 \sin\left(\frac{\sqrt{3}}{2}x\right)\right)$$

Choice (2) is the answer.

4.15. The particular solution of the non-homogeneous second-order differential equation can be solved using inverse operator method. If $r(x) = p_n(x)$, in which $p_n(x)$ is a polynomial of degree n, we can use the following rule.

$$\frac{1}{F(D)}P_n(x) = q_n(D)p_n(x)$$

Herein, to determine $q_n(D)$, $F(D)$ is ordered based on the ascending exponents of D, then the number of one is repeatedly divided by $F(D)$ until the degree of quotient is n.

Therefore, for $y'' + y = x^2$ we have:

$$y_p = \frac{1}{D^2 + 1}x^2$$

Therefore:

$$
\begin{array}{c|c}
1 & 1 + D^2 \\
\underline{1 + D^2} & 1 - D^2 \\
-D^2 & \\
\underline{-D^2 - D^4} & \\
D^4 &
\end{array}
$$

$$\Rightarrow y_p = \left(1 - D^2\right)x^2 \Rightarrow y_p = x^2 - 2$$

Choice (4) is the answer.

4.16. First, we need to determine the roots of the characteristic equation of differential equation as follows.

$$y'' - 2y' + 2y = 2x - 2 \Rightarrow \lambda^2 - 2\lambda + 2 = 0 \Rightarrow \lambda = \frac{2 \pm \sqrt{4 - 8}}{2} = \frac{2 \pm 2i}{2} = 1 \pm i$$

Therefore, the general solution of the homogeneous differential equation is as follows.

$$y_h = c_1 e^x \cos x + c_2 e^x \sin x$$

The particular solution of the non-homogeneous second-order differential equation can be solved using inverse operator method. If $r(x) = p_n(x)$, in which $p_n(x)$ is a polynomial of degree n, we can use the following rule.

$$\frac{1}{F(D)}P_n(x) = q_n(D)p_n(x)$$

Herein, to determine $q_n(D)$, $F(D)$ is ordered based on the ascending exponents of D, then the number of one is repeatedly divided by $F(D)$ until the degree of quotient is n.

Therefore, for $y'' - 2y' + 2y = 2x - 2$ we have:

$$y_p = \frac{1}{D^2 - 2D + 2}(2x - 2)$$

Therefore:

$$\begin{array}{c|c} 1 & 2 - 2D + D^2 \\ \hline 1 - D + \frac{1}{2}D^2 & \frac{1}{2} + \frac{1}{2}D \\ \hline D - \frac{1}{2}D^2 & \end{array}$$

$$\Rightarrow y_p = \left(\frac{1}{2} + \frac{1}{2}D\right)(2x - 2)$$

Note that D is the derivative operator, therefore:

$$y_p = x$$

Finally, the general solution of the non-homogeneous second-order differential equation is:

$$y = y_h + y_p = e^x(c_1 \cos x + c_2 \sin x) + x$$

If we assume $c_1 = A \sin \alpha$ and $c_2 = A \cos \alpha$, the final answer is:

$$y = y_h + y_p = e^x(A \sin \alpha \cos x + A \cos \alpha \sin x) + x$$

$$\Rightarrow y = Ae^x \sin(x + \alpha) + x$$

Choice (3) is the answer.

4.17. The particular solution of the non-homogeneous second-order differential equation can be solved using inverse operator method. If $r(x) = u(x)e^{\alpha x}$, we can use the following rule.

$$\frac{1}{F(D)}u(x)e^{\alpha x} = e^{\alpha x}\frac{1}{F(D + \alpha)}u(x)$$

Therefore, for $y'' - y = 8xe^x$ we have:

$$y_p = \frac{1}{D^2 - 1}8xe^x = 8e^x\frac{1}{(D + 1)^2 - 1}x = 8e^x\frac{1}{(D + 2)D}x$$

Note that $\frac{1}{D}$ is the integral operator, therefore:

$$y_p = 4e^x\frac{1}{D + 2}x^2$$

Now, we need to divide the number of one by $D + 2$ as follows.

$$\begin{array}{r|l}
1 & 2+D \\
1+\frac{1}{2}D & \frac{1}{2}-\frac{1}{4}D+\frac{1}{8}D^2 \\
\hline
-\frac{1}{2}D & \\
-\frac{1}{2}D-\frac{1}{4}D^2 & \\
\hline
\frac{1}{4}D^2 &
\end{array}$$

$$y_p = 4e^x\left(\frac{1}{2}-\frac{1}{4}D+\frac{1}{8}D^2\right)x^2 = 4e^x\left(\frac{1}{2}x^2-\frac{1}{2}x+\frac{1}{4}\right) \Rightarrow y = \left(1-2x+2x^2\right)e^x$$

Choice (4) is the answer.

4.18. The particular solution of the non-homogeneous second-order differential equation can be solved using inverse operator method. If $r(x) = xu(x)$, we can use the following rule.

$$\frac{1}{F(D)}xu(x) = x\frac{1}{F(D)}u(x) - \frac{F'(D)}{(F(D))^2}u(x)$$

Therefore, for $y'' + 2y' + 3y = x\cos x$ we have:

$$y_p = \frac{1}{D^2+2D+3}x\cos x = x\frac{1}{D^2+2D+3}\cos x - \frac{2D+2}{\left(D^2+2D+3\right)^2}\cos x$$

$$= x\frac{1}{-1+2D+3}\cos x - \frac{2D+2}{(-1+2D+3)^2}\cos x = \frac{x}{2}\frac{1}{D+1}\cos x - \frac{D+1}{2(D+1)^2}\cos x$$

$$= \frac{1}{2}(x-1)\frac{1}{D+1}\cos x = \frac{1}{2}(x-1)\frac{D-1}{D^2-1}\cos x = \frac{1}{2}(x-1)\frac{D-1}{-1-1}\cos x = \frac{-1}{4}(x-1)(-\sin x-\cos x)$$

$$\Rightarrow y_p = \frac{1}{4}(x-1)(\sin x+\cos x)$$

Choice (1) is the answer.

4.19. First, we need to determine the roots of the characteristic equation of differential equation as follows.

$$\left(D^2-1\right)y = 8xe^x \tag{1}$$

$$\Rightarrow \lambda^2-1=0 \Rightarrow \lambda = \pm 1 \tag{2}$$

Therefore, the general solution of the homogeneous differential equation is as follows.

$$y_h = c_1e^{-x}+c_2e^x \tag{3}$$

The particular solution of the non-homogeneous second-order differential equation can be solved using undetermined coefficient method. Herein, first, since $r(x) = 8xe^x$, a general guess is made for y_p as follows.

$$y_p = e^x(Ax+B) \tag{4}$$

Since there is a common term in y_p and y_h (e^x), y_p needs to be multiplied by x^m, where m is the minimum natural number that removes the similarity between the terms of y_p and y_h. Herein, $m = 1$ satisfies the criterion. Therefore:

$$y_p = e^x(Ax + B)x = Ax^2e^x + Bxe^x \tag{5}$$

$$y'_p = 2Axe^x + Ax^2e^x + Be^x + Bxe^x \tag{6}$$

$$y''_p = 2Ae^x + 4Axe^x + Ax^2e^x + 2Be^x + Bxe^x \tag{7}$$

Solving (1), (5), and (7):

$$y'' - y = 8xe^x \Rightarrow 2Ae^x + 4Axe^x + 2Be^x = 8xe^x \Rightarrow (2A + 2B)e^x + 4Axe^x = 8xe^x$$

$$\Rightarrow \begin{cases} 4A = 8 \\ 2A + 2B = 0 \end{cases} \Rightarrow A = 2, B = -2 \tag{8}$$

Solving (5) and (8):

$$\Rightarrow y_p = 2x^2e^x - 2xe^x$$

Finally, the general solution of the non-homogeneous second-order differential equation is:

$$y = y_h + y_p = c_1e^{-x} + c_2e^x + 2x^2e^x - 2xe^x$$

$$y = c_1e^{-x} + e^x(c_2 + 2x^2 - 2x)$$

Choice (4) is the answer.

4.20. First, we need to determine the roots of the characteristic equation of differential equation as follows.

$$y'' - 4y' + 4y = \frac{e^{2x}}{x} \Rightarrow \lambda^2 - 4\lambda + 4 = 0 \Rightarrow (\lambda - 2)^2 = 0 \Rightarrow \lambda = 2, 2$$

Therefore, the general solution of the homogeneous differential equation is as follows.

$$y_h = c_1e^{2x} + c_2xe^{2x} = (c_1 + c_2x)e^{2x}$$

To find the particular solution of the non-homogeneous second-order differential equation, we can use the Lagrange method as follows.

The Wronskian of $y_1 = c_1e^{2x}$ and $y_2 = c_2xe^{2x}$ can be determined as follows.

$$W(x) = \begin{vmatrix} y_1 & y_1 \\ y'_1 & y'_2 \end{vmatrix} = \begin{vmatrix} e^{2x} & xe^{2x} \\ 2e^{2x} & e^{2x} + 2xe^{2x} \end{vmatrix} = e^{4x}$$

Then, by assuming $y''(x) + p(x)y' + q(x)y = r(x)$ as the general form of non-homogeneous second-order differential equation, we have:

$$u_1 = -\int \frac{y_2 r(x)}{W(x)} dx = -\int \frac{xe^{2x}\left(\frac{e^{2x}}{x}\right)}{e^{4x}} dx = -\int dx = -x$$

$$u_2 = \int \frac{y_1 r(x)}{W(x)} dx = \int \frac{e^{2x}\left(\frac{e^{2x}}{x}\right)}{e^{4x}} dx = \int \frac{1}{x} dx = \ln x$$

Therefore, the particular solution of the non-homogeneous second-order differential equation is:

$$y_p = u_1 y_1 + u_2 y_2 = -xe^{2x} + x \ln x\, e^{2x}$$

Finally, the general solution of the non-homogeneous second-order differential equation is:

$$y = y_h + y_p = (c_1 + c_2 x + x \ln x - x)e^{2x}$$

Choice (4) is the answer.

4.21. Based on the information given in the problem, we have:

$$y'' + 6y' + 9y = e^{-3x} \cosh x \tag{1}$$

$$y(0) = 1 \tag{2}$$

$$y(1) = 0 \tag{3}$$

First, we need to determine the roots of the characteristic equation of differential equation as follows.

$$\lambda + 6\lambda + 9 = 0 \Rightarrow (\lambda + 3)^2 = 0 \Rightarrow \lambda = -3, \, -3$$

Therefore, the general solution of the homogeneous differential equation is as follows.

$$y_h = c_1 e^{-3x} + c_2 x e^{-3x}$$

The particular solution of the non-homogeneous second-order differential equation can be solved using inverse operator method. If $r(x) = Ae^{\alpha x}$, we can use the following rule.

$$y_p = \frac{1}{F(D)} e^{\alpha x} = \begin{cases} \dfrac{1}{F(\alpha)} e^{\alpha x} & F(\alpha) \neq 0 \\[2mm] \dfrac{x}{F'(\alpha)} e^{\alpha x} & F(\alpha) = 0, F'(\alpha) \neq 0 \\[2mm] \dfrac{x^2}{F''(\alpha)} e^{\alpha x} & F(\alpha) = F'(\alpha) = 0, F''(\alpha) \neq 0 \\[2mm] \vdots & \\[2mm] \dfrac{x^k}{F^{(k)}(\alpha)} e^{\alpha x} & F(\alpha) = F'(\alpha) = \cdots = F^{(k-1)}(\alpha) = 0, F^{(k)}(\alpha) \neq 0 \end{cases}$$

Therefore, for $y'' + 6y' + 9y = e^{-3x} \cosh x$, we have:

$$y_p = \frac{1}{D^2 + 6D + 9} e^{-3x} \cosh x = \frac{1}{D^2 + 6D + 9} e^{-3x}\left(\frac{e^x + e^{-x}}{2}\right) = \frac{1}{D^2 + 6D + 9}\left(\frac{e^{-2x} + e^{-4x}}{2}\right)$$

$$y_p = \frac{1}{2} \frac{1}{D^2 + 6D + 9} e^{-2x} + \frac{1}{2} \frac{1}{D^2 + 6D + 9} e^{-4x} = \frac{e^{-2x} + e^{-4x}}{2}$$

Finally, the general solution of the non-homogeneous second-order differential equation is:

$$y = y_h + y_p = c_1 e^{-3x} + c_2 x e^{-3x} + \frac{e^{-2x} + e^{-4x}}{2} \qquad (4)$$

Solving (2) and (4):

$$c_1 + \frac{1+1}{2} = 1 \Rightarrow c_1 = 0 \qquad (5)$$

Solving (3)–(5):

$$0 + c_2 e^{-3} + \frac{e^{-2} + e^{-4}}{2} = 0 \Rightarrow c_2 = -\frac{e + e^{-1}}{2} = -\cosh(1) \qquad (6)$$

Solving (4)–(6):

$$y = -\cosh(1) x e^{-3x} + \frac{e^{-2x} + e^{-4x}}{2}$$

Therefore, for $x = -1$, we have:

$$y(-1) = \cosh(1) e^3 + \frac{e^2 + e^4}{2} = e^3 \cosh(1) + e^3 \frac{e + e^{-1}}{2} = 2e^3 \cosh(1)$$

Choice (4) is the answer.

4.22. Based on the information given in the problem, we have:

$$y(x = 0) = 1 \qquad (1)$$

$$y'(x = 0) = 1 \qquad (2)$$

$$y'' + 4y' = \sin 2x \qquad (3)$$

First, we need to determine the roots of the characteristic equation of differential equation as follows.

$$\Rightarrow \lambda^2 + 4 = 0 \Rightarrow \lambda = \pm 2i \qquad (4)$$

Therefore, the general solution of the homogeneous differential equation is as follows.

$$y_h = c_1 \cos(2x) + c_2 \sin(2x) \qquad (5)$$

The particular solution of the non-homogeneous second-order differential equation can be solved using undetermined coefficient method. Herein, since $r(x) = \sin 2x$, a general guess is made for y_p as follows.

$$y_p = A \cos(2x) + B \sin(2x) \qquad (6)$$

Since there is a common term in y_p and y_h ($\sin 2x$ and $\cos 2x$), y_p needs to be multiplied by x^m, where m is the minimum natural number that removes the similarity between the terms of y_p and y_h. Herein, $m = 1$ satisfies the criterion. Therefore:

$$\Rightarrow y_p = Ax \cos(2x) + Bx \sin(2x) \qquad (7)$$

$$y'_p = A\cos(2x) - 2Ax\sin(2x) + B\sin(2x) + 2Bx\cos(2x) \tag{8}$$

$$y''_p = -4A\sin(2x) - 4Ax\cos(2x) + 4B\cos(2x) - 4Bx\sin(2x) \tag{9}$$

Solving (3), (8), and (9):

$$-4A\sin(2x) + 4B\cos(2x) = \sin 2x \Rightarrow \begin{cases} -4A = 1 \Rightarrow A = -\dfrac{1}{4} \\ 4B = 0 \Rightarrow B = 0 \end{cases} \tag{10}$$

Solving (7) and (10):

$$y_p = -\frac{x}{4}\cos(2x)$$

Therefore, the general solution of the non-homogeneous differential equation is as follows.

$$y = y_h + y_p = c_1\cos(2x) + c_2\sin(2x) - \frac{x}{4}\cos(2x) \tag{11}$$

$$\Rightarrow y' = -2c_1\sin(2x) + 2c_2\cos(2x) - \frac{1}{4}\cos(2x) + \frac{x}{2}\sin(2x) \tag{12}$$

Solving (1) and (11):

$$c_1 + 0 - 0 = 1 \Rightarrow c_1 = 1 \tag{13}$$

Solving (2) and (12):

$$0 + 2c_2 - \frac{1}{4} + 0 = 1 \Rightarrow c_2 = \frac{5}{8} \tag{14}$$

Solving (11), (13), and (14):

$$y = \cos(2x) + \frac{5}{8}\sin(2x) - \frac{x}{4}\cos(2x)$$

$$\Rightarrow y\left(\frac{\pi}{4}\right) = \frac{5}{8}$$

Choice (2) is the answer.

4.23. Based on the information given in the problem, we have:

$$y(x = 0) = 1 \tag{1}$$

$$y'(x = 0) = -1 \tag{2}$$

$$y''(x = 0) = 1 \tag{3}$$

$$y'''(x = 0) = 0 \tag{4}$$

$$y^{(4)} + 5y'' + 4y = 0 \tag{5}$$

First, we need to determine the roots of the characteristic equation of differential equation as follows.

$$\Rightarrow \lambda^4 + 5\lambda^2 + 4 = 0 \Rightarrow (\lambda^2 + 1)(\lambda^2 + 4) = 0 \Rightarrow \begin{cases} \lambda^2 + 1 = 0 \Rightarrow \lambda = \pm i \\ \lambda^2 + 4 = 0 \Rightarrow \lambda = \pm 2i \end{cases} \tag{6}$$

Therefore, the general solution of the homogeneous differential equation is as follows.

$$y = c_1 \cos x + c_2 \sin x + c_3 \cos(2x) + c_4 \sin(2x) \tag{7}$$

$$y' = -c_1 \sin x + c_2 \cos x - 2c_3 \sin(2x) + 2c_4 \cos(2x) \tag{8}$$

$$y'' = -c_1 \cos x - c_2 \sin x - 4c_3 \cos(2x) - 4c_4 \sin(2x) \tag{9}$$

$$y''' = c_1 \sin x - c_2 \cos x + 8c_3 \sin(2x) - 8c_4 \cos(2x) \tag{10}$$

Solving (1) and (7):

$$c_1 + c_3 = 1 \tag{11}$$

Solving (2) and (8):

$$c_2 + 2c_4 = -1 \tag{12}$$

Solving (3) and (9):

$$-c_1 - 4c_3 = 1 \tag{13}$$

Solving (4) and (10):

$$-c_2 - 8c_4 = 0 \tag{14}$$

Solving (11)–(14):

$$c_1 = \frac{5}{3}, c_2 = \frac{-4}{3}, c_3 = \frac{-2}{3}, c_4 = \frac{1}{6} \tag{15}$$

Solving (7) and (15):

$$y = \frac{5}{3} \cos x - \frac{4}{3} \sin x - \frac{2}{3} \cos(2x) + \frac{1}{6} \sin(2x)$$

$$\Rightarrow y(\pi) = \frac{-7}{3}$$

Choice (2) is the answer.

Problems: Series and Their Applications in Solving Differential Equations

5

Abstract

In this chapter, differential equations including variable coefficients are solved by using series. In this chapter, the problems are categorized in different levels based on their difficulty levels (easy, normal, and hard) and calculation amounts (small, normal, and large). Additionally, the problems are ordered from the easiest problem with the smallest computations to the most difficult problems with the largest calculations.

5.1. What is the general solution of the differential equation below?

$$2x^2 y'' + xy' - (x+1)y = 0$$

Difficulty level   ○ Easy   ● Normal   ○ Hard
Calculation amount   ○ Small   ● Normal   ○ Large

1) $c_1 x^{\frac{1}{3}} \sum_{n=0}^{\infty} a_n x^n + c_2 \sum_{n=0}^{\infty} b_n x^n$

2) $c_1 x \sum_{n=0}^{\infty} a_n x^n + c_2 \ln x \sum_{n=0}^{\infty} b_n x^n$

3) $c_1 x \sum_{n=0}^{\infty} a_n x^n + c_2 x^{-\frac{1}{2}} \sum_{n=0}^{\infty} b_n x^n$

4) $c_1 x^{-1} \sum_{n=0}^{\infty} a_n x^n + c_2 x^{\frac{1}{2}} \sum_{n=0}^{\infty} b_n x^n$

5.2. Which one of the choices is correct about the differential equation below?

$$(x+1)xy'' - (2x+1)y' + 2y = 0$$

Difficulty level   ○ Easy   ● Normal   ○ Hard
Calculation amount   ○ Small   ● Normal   ○ Large

1) Both $x = 0$ and $x = -1$ are regular singular points.
2) Both $x = 0$ and $x = -1$ are irregular singular points.
3) $x = 0$ is a regular singular point and $x = 1$ is an irregular singular point.
4) $x = 0$ is an irregular singular point and $x = 1$ is a regular singular point.

5.3. Which one of the choices is correct about the following differential equation?

$$x^2(1-x)y'' + y' - y = 0$$

Difficulty level   ○ Easy   ● Normal   ○ Hard
Calculation amount   ○ Small   ● Normal   ○ Large

1) Both $x = 0$ and $x = 1$ are regular singular points.
2) Both $x = 0$ and $x = 1$ are irregular singular points.
3) $x = 0$ is a regular singular point and $x = 1$ is an irregular singular point.
4) $x = 0$ is an irregular singular point and $x = 1$ is a regular singular point.

5.4. Determine the characteristic equation of the differential equation below.

$$x^2 y'' + \left(x^2 + \frac{5}{36}\right) y = 0$$

Difficulty level   ○ Easy   ● Normal   ○ Hard
Calculation amount   ○ Small   ● Normal   ○ Large

1) $r^2 - r + \frac{5}{36} = 0$
2) $r^2 + \frac{5}{36} r - 1 = 0$
3) $r^2 - \frac{5}{36} r + 1 = 0$
4) $r^2 + r - \frac{5}{36} = 0$

5.5. What are the two independent solutions of the differential equation below?

$$3xy'' + 2y' + y = 0$$

Difficulty level   ○ Easy   ● Normal   ○ Hard
Calculation amount   ○ Small   ● Normal   ○ Large

1) $x^{\frac{1}{3}} \sum_{n=0}^{\infty} a_n x^n$ and $\sum_{n=0}^{\infty} b_n x^n$
2) $x^{-1} \sum_{n=0}^{\infty} a_n x^n$ and $\sum_{n=0}^{\infty} b_n x^n$
3) $x^{-2} \sum_{n=0}^{\infty} a_n x^n$ and $x^{\frac{1}{3}} \sum_{n=0}^{\infty} b_n x^n$
4) $\sum_{n=0}^{\infty} a_n x^n$ and $\ln x \sum_{n=0}^{\infty} a_n x^n + \sum_{n=0}^{\infty} b_n x^n$

5.6. Which one of the choices is correct about the differential equation below?

$$y'' - xy' = 0, \quad y = \sum_{n=0}^{\infty} a_n x^n$$

Difficulty level   ○ Easy   ○ Normal   ● Hard
Calculation amount   ○ Small   ● Normal   ○ Large

1) $a_{n+2} = \frac{a_n}{(n+1)(n+2)}, n \geq 0$
2) $a_{n+2} = \frac{na_n}{(n+1)(n+2)}, n \geq 0$
3) $a_{n+3} = \frac{-a_n}{(n+1)(n+2)}, n \geq 0$
4) $a_{n+3} = \frac{a_n}{(n+1)(n+3)}, n \geq 0$

5.7. Calculate the Maclaurin series expansion of the following differential equation with the given primary conditions.

$$y''(x) + y(x) + y^3(x) = \cos x, \quad y(x = 0) = 0, y'(x = 0) = 1$$

Difficulty level   ○ Easy   ○ Normal   ● Hard
Calculation amount   ○ Small   ● Normal   ○ Large

1) $x + \frac{x^2}{2!} + \frac{x^3}{3!} + \cdots$

2) $x - \frac{x^2}{2!} + \frac{x^3}{3!} + \cdots$

3) $x + \frac{x^2}{2} + \frac{x^3}{3} + \cdots$

4) $x + \frac{x^2}{2!} - \frac{x^3}{3!} + \cdots$

5.8. What is the general solution of the differential equation below based on Bessel functions?

$$xy'' - y' + xy = 0$$

Difficulty level ○ Easy ○ Normal ● Hard
Calculation amount ○ Small ● Normal ○ Large

1) $x(c_1 J_1(x) + c_2 Y_1(x))$

2) $x^2(c_1 J_1(x) + c_2 Y_1(x))$

3) $x(c_1 J_1(x) + c_2 J_{-1}(x))$

4) $x^2(c_1 J_1(x) + c_2 J_{-1}(x))$

5.9. Which one of the choices is correct about the differential equation below?

$$y'' + x^2 y = 0, \quad y = \sum_{n=0}^{\infty} a_n x^n$$

Difficulty level ○ Easy ○ Normal ● Hard
Calculation amount ○ Small ● Normal ○ Large

1) $a_{n+2} = \frac{a_n}{(n+1)(n+2)}, n \geq 0$

2) $a_{n+3} = \frac{a_{n+1}}{(n+3)(n+4)}, n \geq 0$

3) $a_{n+3} = -\frac{a_{n-1}}{(n+3)(n+4)}, n \geq 0$

4) $a_{n+4} = -\frac{a_n}{(n+3)(n+4)}, n \geq 0$

Solutions of Problems: Series and Their Applications in Solving Differential Equations

6

Abstract

In this chapter, the problems of the fifth chapter are fully solved, in detail, step-by-step, and with different methods.

6.1. Consider the general form of the homogeneous second-order differential equation including variable coefficients as follows. Herein, consider that $a_2(x)$, $a_1(x)$, and $a_0(x)$ are analytic at $x = x_0$.

$$a_2(x)y''(x) + a_1(x)y'(x) + a_0(x) = 0 \tag{1}$$

$$\xrightarrow{\times \dfrac{1}{a_2(x)}} y''(x) + \frac{a_1(x)}{a_2(x)}y'(x) + \frac{a_0(x)}{a_2(x)} = 0 \tag{2}$$

Case 1: $x = x_0$ is an ordinary (nonsingular) point if $a_2(x_0) \neq 0$. Then, the general solution of the homogeneous second-order differential equation is as follows.

$$y(x) = \sum_{n=0}^{\infty} a_n(x - x_0)^n \tag{3}$$

Case 2: $x = x_0$ is a regular singular point if $a_2(x_0) = 0$, but both limits, presented in (4) and (5), are available.

$$p_0 = \lim_{x \to x_0} (x - x_0)\frac{a_1(x)}{a_2(x)} \tag{4}$$

$$q_0 = \lim_{x \to x_0} (x - x_0)^2 \frac{a_0(x)}{a_2(x)} \tag{5}$$

In this case, first, we need to determine the roots of the characteristic equation of differential equation as follows.

$$r^2 + (p_0 - 1)r + q_0 = 0 \tag{6}$$

Then, the general solution of the homogeneous second-order differential equation is as follows.

$$y(x) = c_1 y_1(x) + c_2 y_2(x) \tag{7}$$

where

$$y_1(x) = (x - x_0)^{r_1} \sum_{n=0}^{\infty} a_n (x - x_0)^n, \quad if \ r_1 \geq r_2 \tag{8}$$

If $r_1 - r_2 \notin \mathbb{Z}$:

$$y_2(x) = (x - x_0)^{r_2} \sum_{n=0}^{\infty} b_n (x - x_0)^n \tag{9}$$

If $r_1 = r_2$:

$$y_2(x) = y_1(x) \ln |x - x_0| + (x - x_0)^{r_1} \sum_{n=0}^{\infty} b_n (x - x_0)^n \tag{10}$$

If $r_1 - r_2 \in \mathbb{Z} - \{0\}$:

$$y_2(x) = \alpha y_1(x) \ln |x - x_0| + (x - x_0)^{r_2} \sum_{n=0}^{\infty} b_n (x - x_0)^n, b_0 \neq 0 \tag{11}$$

Case 3: $x = x_0$ is an irregular singular point if $a_2(x_0) = 0$, and one of the limits, presented in (4) and (5), is unavailable.

Based on the information given in the problem, we have:

$$2x^2 y'' + xy' - (x + 1)y = 0 \Rightarrow y'' + \frac{x}{2x^2} y' - \frac{(x + 1)}{2x^2} y = 0$$

Since $a_2(0) = 2x^2|_{x=0} = 0$, $x = 0$ is a singular point. Moreover, since both limits below are available, $x = 0$ is a regular singular point.

$$p_0 = \lim_{x \to x_0} (x - x_0) \frac{a_1(x)}{a_2(x)} = \lim_{x \to 0} x \frac{x}{2x^2} = \frac{1}{2}$$

$$q_0 = \lim_{x \to x_0} (x - x_0)^2 \frac{a_0(x)}{a_2(x)} = \lim_{x \to 0} x^2 \frac{-(x + 1)}{2x^2} = -\frac{1}{2}$$

Now, we need to determine the roots of the characteristic equation of differential equation as follows.

$$r^2 + (p_0 - 1)r + q_0 = 0 \Rightarrow r^2 + \left(\frac{1}{2} - 1\right)r + \left(-\frac{1}{2}\right) = 0 \Rightarrow r^2 - \frac{1}{2}r - \frac{1}{2} = 0 \Rightarrow r_{1,2} = -\frac{1}{2}, 1$$

Since $r_1 - r_2 \notin \mathbb{Z}$, we have:

$$y_1 = c_1 x^1 \sum_{n=0}^{\infty} a_n x^n, \quad y_2 = c_2 x^{-\frac{1}{2}} \sum_{n=0}^{\infty} b_n x^n$$

Choice (3) is the answer.

6.2. Based on the information given in the problem, we have:

$$(x + 1)xy'' - (2x + 1)y' + 2y = 0 \Rightarrow y'' - \frac{(2x + 1)}{(x + 1)x} y' + \frac{2}{(x + 1)x} y = 0$$

Since $a_2(0) = (x + 1)x|_{x=0} = 0$ and $a_2(-1) = (x + 1)x|_{x=-1} = 0$, $x = 0$ and $x = -1$ are singular points.

Since both limits below are available for $x = 0$, the point is a regular singular point.

$$p_0 = \lim_{x \to x_0} (x - x_0) \frac{a_1(x)}{a_2(x)} = \lim_{x \to 0} x \left(-\frac{(2x+1)}{(x+1)x} \right) = -1$$

$$q_0 = \lim_{x \to x_0} (x - x_0)^2 \frac{a_0(x)}{a_2(x)} = \lim_{x \to 0} x^2 \frac{2}{(x+1)x} = 0$$

Likewise, since both limits below are available for $x = -1$, the point is a regular singular point.

$$p_0 = \lim_{x \to x_0} (x - x_0) \frac{a_1(x)}{a_2(x)} = \lim_{x \to -1} (x+1) \left(-\frac{(2x+1)}{(x+1)x} \right) = -1$$

$$q_0 = \lim_{x \to x_0} (x - x_0)^2 \frac{a_0(x)}{a_2(x)} = \lim_{x \to -1} (x+1)^2 \frac{2}{(x+1)x} = 0$$

Choice (1) is the answer.

6.3. Based on the information given in the problem, we have:

$$x^2(1-x)y'' + y' - y = 0 \Rightarrow y'' + \frac{1}{x^2(1-x)} y' - \frac{1}{x^2(1-x)} y = 0$$

Since $a_2(0) = x^2(1-x)|_{x=0} = 0$ and $a_2(1) = x^2(1-x)|_{x=1} = 0$, $x = 0$ and $x = 1$ are singular points.

Since one of the limits below is unavailable for $x = 0$, the point is an irregular singular point.

$$p_0 = \lim_{x \to x_0} (x - x_0) \frac{a_1(x)}{a_2(x)} = \lim_{x \to 0} x \frac{1}{x^2(1-x)} = \lim_{x \to 0} \frac{1}{x(1-x)}$$

$$q_0 = \lim_{x \to x_0} (x - x_0)^2 \frac{a_0(x)}{a_2(x)} = \lim_{x \to 0} x^2 \left(-\frac{1}{x^2(1-x)} \right) = -1$$

Likewise, since both limits below are available for $x = 1$, the point is a regular singular point.

$$p_0 = \lim_{x \to x_0} (x - x_0) \frac{a_1(x)}{a_2(x)} = \lim_{x \to 1} (x-1) \frac{1}{x^2(1-x)} = -1$$

$$q_0 = \lim_{x \to x_0} (x - x_0)^2 \frac{a_0(x)}{a_2(x)} = \lim_{x \to 1} (x-1)^2 \left(-\frac{1}{x^2(1-x)} \right) = 0$$

Choice (4) is the answer.

6.4. Based on the information given in the problem, we have:

$$x^2 y'' + \left(x^2 + \frac{5}{36} \right) y = 0 \Rightarrow y'' + \left(\frac{x^2 + \frac{5}{36}}{x^2} \right) y = 0 \tag{1}$$

The characteristic equation of the differential equation can be determined as follows.

$$r^2 + (p_0 - 1)r + q_0 = 0 \tag{2}$$

where

$$p_0 = \lim_{x \to x_0}(x - x_0)\frac{a_1(x)}{a_2(x)} = \lim_{x \to 0}x \times 0 = 0 \qquad (3)$$

$$q_0 = \lim_{x \to x_0}(x - x_0)^2\frac{a_0(x)}{a_2(x)} = \lim_{x \to 0}x^2\left(\frac{x^2 + \frac{5}{36}}{x^2}\right) = \frac{5}{36} \qquad (4)$$

Solving (2)–(4):

$$r^2 + (0 - 1)r + \frac{5}{36} = 0 \Rightarrow r^2 - r + \frac{5}{36} = 0$$

Choice (1) is the answer.

6.5. Based on the information given in the problem, we have:

$$3xy'' + 2y' + y = 0 \Rightarrow y'' + \frac{2}{3x}y' + \frac{1}{3x}y = 0$$

Since $a_2(0) = 3x|_{x=0} = 0$ but both limits below are available, $x = 0$ is a regular singular point.

$$p_0 = \lim_{x \to x_0}(x - x_0)\frac{a_1(x)}{a_2(x)} = \lim_{x \to 0}x\frac{2}{3x} = \frac{2}{3}$$

$$q_0 = \lim_{x \to x_0}(x - x_0)^2\frac{a_0(x)}{a_2(x)} = \lim_{x \to 0}x^2\frac{1}{3x} = 0$$

Now, we need to determine the roots of the characteristic equation of the differential equation as follows.

$$r^2 + (p_0 - 1)r + q_0 = 0 \Rightarrow r^2 + \left(\frac{2}{3} - 1\right)r + 0 = 0 \Rightarrow r^2 - \frac{1}{3}r = 0 \Rightarrow r_{1,2} = 0, \frac{1}{3}$$

Since $r_1 - r_2 \notin \mathbb{Z}$, we have:

$$y_1 = x^{\frac{1}{3}}\sum_{n=0}^{\infty}a_n x^n, \quad y_2 = \sum_{n=0}^{\infty}b_n x^n$$

Choice (1) is the answer.

6.6. Based on the information given in the problem, we have:

$$y'' - xy' = 0 \qquad (1)$$

$$y = \sum_{n=0}^{\infty}a_n x^n \qquad (2)$$

Therefore:

$$\Rightarrow y' = \sum_{n=0}^{\infty}na_n x^{n-1} \qquad (3)$$

$$\Rightarrow y'' = \sum_{n=0}^{\infty} n(n-1)a_n x^{n-2} \tag{4}$$

Solving (1), (3), and (4):

$$\sum_{n=0}^{\infty} n(n-1)a_n x^{n-2} - \sum_{n=0}^{\infty} na_n x^n = 0$$

$$\xrightarrow{\;n \to n+2 \text{ in the first series}\;} \sum_{n=-2}^{\infty} (n+2)(n+1)a_{n+2} x^n - \sum_{n=0}^{\infty} na_n x^n = 0$$

$$\Rightarrow \left[0 + 0 + \sum_{n=0}^{\infty} (n+2)(n+1)a_{n+2} x^n \right] - \sum_{n=0}^{\infty} na_n x^n = 0$$

$$\Rightarrow \sum_{n=0}^{\infty} [(n+2)(n+1)a_{n+2} - na_n]x^n = 0 \Rightarrow (n+2)(n+1)a_{n+2} - na_n = 0, \quad n \geq 0$$

$$\Rightarrow a_{n+2} = \frac{na_n}{(n+2)(n+1)}, \quad n \geq 0$$

Choice (2) is the answer.

6.7. The Maclaurin series expansion of the function of $y(x)$ can be determined as follows.

$$y(x) = y(0) + y'(0)x + \frac{y''(0)}{2!}x^2 + \frac{y'''(0)}{3!}x^3 + \cdots \tag{1}$$

Based on the information given in the problem, we have:

$$y''(x) + y(x) + y^3(x) = \cos x \tag{2}$$

$$y(x = 0) = 0 \tag{3}$$

$$y'(x = 0) = 1 \tag{4}$$

Solving Equation (2) for $x = 0$ results in the following term.

$$y''(0) + y(0) + (y(0))^3 = \cos 0 \tag{5}$$

Solving (3)–(5):

$$\Rightarrow y''(0) + 0 + 0 = 1 \Rightarrow y''(0) = 1 \tag{6}$$

Calculating the first derivative of (2):

$$y''' + y' + 3y'y^2 = -\sin x \tag{7}$$

Solving equation (7) for $x = 0$:

$$y'''(0) + y'(0) + 3y'(0)(y(0))^2 = -\sin 0 \tag{8}$$

Solving (3), (4), and (8):

$$y'''(0) + 1 + 3(1)(0) = 0 \Rightarrow y'''(0) = -1 \tag{9}$$

Solving (1), (3), (4), (6), and (9):

$$\Rightarrow y(x) = 0 + x + \frac{1}{2!}x^2 + \frac{-1}{3!}x^3 + \cdots$$

$$\Rightarrow y(x) = x + \frac{1}{2!}x^2 - \frac{1}{3!}x^3 + \cdots$$

Choice (4) is the answer.

6.8. Based on the information given in the problem, we have:

$$xy'' - y' + xy = 0 \tag{1}$$

Defining a new variable:

$$y \triangleq xu \tag{2}$$

Therefore:

$$y' = u + xu' \tag{3}$$

$$y'' = 2u' + xu'' \tag{4}$$

Solving (1)–(4):

$$x(xu'' + 2u') - (u + xu') + x(xu) = 0$$

$$\Rightarrow x^2u'' + xu' + (x^2 - 1)u = 0 \tag{5}$$

Equation (5) is a first-order Bessel differential equation with respect to u. Therefore, its general solution is as follows.

$$u(x) = c_1 J_1(x) + c_2 Y_1(x) \tag{6}$$

Solving (2) and (6):

$$y(x) = x(c_1 J_1(x) + c_2 Y_1(x))$$

Choice (1) is the answer.

6.9. Based on the information given in the problem, we have:

$$y'' + x^2 y = 0 \tag{1}$$

$$y = \sum_{n=0}^{\infty} a_n x^n \tag{2}$$

Therefore:

$$y' = \sum_{n=0}^{\infty} n a_n x^{n-1} \tag{3}$$

$$y'' = \sum_{n=0}^{\infty} n(n-1) a_n x^{n-2} \tag{4}$$

Solving (1) and (4):

$$\sum_{n=0}^{\infty} n(n-1) a_n x^{n-2} + \sum_{n=0}^{\infty} a_n x^{n+2} = 0$$

$$\xrightarrow{n \rightarrow n+4 \ in \ the \ first \ series} \sum_{n=-4}^{\infty} (n+4)(n+3) a_{n+4} x^{n+2} + \sum_{n=0}^{\infty} a_n x^{n+2} = 0$$

$$\Rightarrow \left[0 + 0 + 2a_2 + 6a_3 x + \sum_{n=0}^{\infty} (n+4)(n+3) a_{n+4} x^{n+2} \right] + \sum_{n=0}^{\infty} a_n x^{n+2} = 0$$

$$\Rightarrow 2a_2 + 6a_3 x + \sum_{n=0}^{\infty} [(n+4)(n+3) a_{n+4} + a_n] x^n = 0$$

$$\Rightarrow \begin{cases} a_2 = 0 \\ a_3 = 0 \\ (n+4)(n+3) a_{n+4} + a_n = 0, n \geq 0 \Rightarrow a_{n+4} = -\dfrac{a_n}{(n+4)(n+3)}, n \geq 0 \end{cases}$$

Choice (4) is the answer.

Problems: Laplace Transform and Its Applications in Solving Differential Equations

7

Abstract

In this chapter, Laplace transform is applied to solve different types of differential equations. In addition, many examples about Laplace transform and inverse Laplace transform of different types of functions are presented. In this chapter, the problems are categorized in different levels based on their difficulty levels (easy, normal, and hard) and calculation amounts (small, normal, and large). Additionally, the problems are ordered from the easiest problem with the smallest computations to the most difficult problems with the largest calculations.

7.1. Determine the inverse Laplace transform of the following term.

$$F(s) = \frac{1}{s^4}$$

Difficulty level   ● Easy   ○ Normal   ○ Hard
Calculation amount   ● Small   ○ Normal   ○ Large

1) $\frac{t}{2}$

2) $\frac{t^2}{4}$

3) $\frac{t^3}{6}$

4) $\frac{t^4}{8}$

7.2. What is the inverse Laplace transform of the term below?

$$F(s) = \frac{1}{s^2 + 25}$$

Difficulty level   ● Easy   ○ Normal   ○ Hard
Calculation amount   ● Small   ○ Normal   ○ Large

1) $\frac{1}{5} \sin 5t$

2) $\frac{1}{5} \cos 5t$

3) $5 \sin 5t$

4) $5 \cos 5t$

7.3. Determine the inverse Laplace transform of the following term.

$$F(s) = \frac{3}{s + 4}$$

Difficulty level   ● Easy   ○ Normal   ○ Hard
Calculation amount   ● Small   ○ Normal   ○ Large

1) $3(t+3)$
2) $4e^{-3t}$
3) $3e^{4t}$
4) $3e^{-4t}$

7.4. Calculate the inverse Laplace transform of the following term.

$$F(s) = \frac{1}{s^2 + 8^2}$$

Difficulty level   ● Easy   ○ Normal   ○ Hard
Calculation amount   ● Small   ○ Normal   ○ Large

1) $8 \cos 8t$
2) $8 \sin 8t$
3) $\frac{1}{8} \cos 8t$
4) $\frac{1}{8} \sin 8t$

7.5. Calculate the Laplace transform of the function below.

$$f(t) = \cos 2t$$

Difficulty level   ● Easy   ○ Normal   ○ Hard
Calculation amount   ● Small   ○ Normal   ○ Large

1) $\frac{2s}{s^2-4}$
2) $\frac{2s}{s^2+4}$
3) $\frac{s}{s^2+4}$
4) $\frac{s}{s^2-4}$

7.6. Calculate the Laplace transform of the following term.

$$f(t) = \frac{\sin 4t}{4}$$

Difficulty level   ● Easy   ○ Normal   ○ Hard
Calculation amount   ● Small   ○ Normal   ○ Large

1) $\frac{4}{s^2+16}$
2) $\frac{1}{s^2+16}$
3) $\frac{4}{s^2-16}$
4) $\frac{1}{s^2-16}$

7.7. Calculate the Laplace transform of the following term.

$$f(t) = \frac{\sinh 5t}{5}$$

Difficulty level   ● Easy   ○ Normal   ○ Hard
Calculation amount   ● Small   ○ Normal   ○ Large

1) $\frac{5}{s^2-25}$
2) $\frac{1}{s^2-25}$
3) $\frac{5}{s^2+25}$
4) $\frac{1}{s^2+25}$

7.8. Determine the Laplace transform of the function below.

$$f(t) = te^t$$

Difficulty level   ○ Easy   ● Normal   ○ Hard
Calculation amount   ● Small   ○ Normal   ○ Large

1) $\frac{1}{(s+1)^2}$

2) $\frac{6}{(s-1)^2}$

3) $\frac{6}{(s+1)^2}$

4) $\frac{1}{(s-1)^2}$

7.9. Determine the Laplace transform of the function below.

$$f(t) = e^t \sin t$$

Difficulty level   ○ Easy   ● Normal   ○ Hard
Calculation amount   ● Small   ○ Normal   ○ Large

1) $\frac{1}{(s+1)^2+1}$

2) $\frac{s}{(s-1)^2+1}$

3) $\frac{s}{(s+1)^2-1}$

4) $\frac{1}{(s-1)^2+1}$

7.10. Calculation the value of the convolution of the functions of $f(t) = t$ and $g(t) = 1$, that is, $f(t) * g(t)$.

Difficulty level   ○ Easy   ● Normal   ○ Hard
Calculation amount   ● Small   ○ Normal   ○ Large

1) $\frac{t^3}{3}$

2) t

3) 1

4) $\frac{t^2}{2}$

7.11. Determine the Laplace transform of the function below.

$$f(t) = 2e^{-2t} \sin 2t$$

Difficulty level   ○ Easy   ● Normal   ○ Hard
Calculation amount   ● Small   ○ Normal   ○ Large

1) $\frac{1}{(s+1)^2+4}$

2) $\frac{2}{(s+1)^2+4}$

3) $\frac{1}{(2s+1)^2+4}$

4) $\frac{4}{(s+2)^2+4}$

7.12. Determine the Laplace transform of the function below.

$$f(t) = t^2 \sin t$$

Difficulty level   ○ Easy   ● Normal   ○ Hard
Calculation amount   ● Small   ○ Normal   ○ Large

1) $\frac{-2s}{(s^2+1)^2}$

2) $\frac{6s^2-2}{(s^2+1)^3}$

3) $\frac{4s^2+2}{(s^2+1)^3}$

4) $\frac{2s^2+1}{(s^2+1)^2}$

7.13. Solve the equation below.

$$f(t) = L^{-1}\left[\frac{1}{(s-2)(s-1)}\right]$$

Difficulty level   ○ Easy   ● Normal   ○ Hard
Calculation amount   ● Small   ○ Normal   ○ Large

1) $f(t) = e^{2t}$
2) $f(t) = e^{t}$
3) $f(t) = e^{2t} - e^{t}$
4) $f(t) = e^{2t} + e^{t}$

7.14. Calculate the Laplace transform of the following term.

$$f(t) = \sin t \cos t$$

Difficulty level   ○ Easy   ● Normal   ○ Hard
Calculation amount   ● Small   ○ Normal   ○ Large

1) $\frac{2}{s^2+4}$

2) $\frac{1}{2(s^2+4)}$

3) $\frac{1}{s^2+4}$

4) $\frac{1}{s^2+3}$

7.15. Calculate the inverse Laplace transform of the following term.

$$F(s) = \frac{1}{s^2(s^2+\omega^2)}$$

Difficulty level   ○ Easy   ● Normal   ○ Hard
Calculation amount   ● Small   ○ Normal   ○ Large

1) $\frac{1}{\omega}\left(1 - \frac{1}{\omega}\sin \omega t\right)$

2) $\frac{1}{\omega^2}\left(t - \frac{1}{\omega}\sin \omega t\right)$

3) $\frac{1}{\omega^2}\left(1 - \sin \omega t\right)$

4) $\frac{1}{\omega^2}\left(t - \frac{1}{\omega^2}\sin \omega t\right)$

7.16. Calculate the Laplace transform of the following term.

$$f(t) = \frac{t \sinh 2t}{4}$$

Difficulty level   ○ Easy   ● Normal   ○ Hard
Calculation amount   ● Small   ○ Normal   ○ Large

1) $\frac{s}{(s^2+4)^2}$

2) $\frac{4s}{(s^2+4)^2}$

3) $\frac{4s}{(s^2-4)^2}$

4) $\frac{s}{(s^2-4)^2}$

7.17. Determine the inverse Laplace transform of the term below.

$$F(s) = \frac{3s+2}{s^2+2s+10}$$

Difficulty level   ○ Easy   ● Normal   ○ Hard
Calculation amount   ○ Small   ● Normal   ○ Large

1) $\frac{1}{3}e^{-t}\cos 3t + 3e^t \sin 3t$

2) $e^t \cos 3t + 3e^t \sin 3t$

3) $3e^{-t}\cos 3t - \frac{1}{3}e^{-t}\sin 3t$

4) $3e^{-t}\cos 3t + e^{-t}\sin 3t$

7.18. Calculate the value of the following term.

$$f(t) = L^{-1}\left(\frac{1}{S^2+3s+\frac{13}{4}}\right)$$

Difficulty level   ○ Easy   ● Normal   ○ Hard
Calculation amount   ○ Small   ● Normal   ○ Large

1) $e^{-t}\sin \frac{3}{2}t$

2) $e^{-\frac{3}{2}t}\sin t$

3) $e^{-\frac{3}{2}\cos t}$

4) $te^{-\frac{3}{2}t}$

7.19. Solve the equation below using Laplace transform.

$$z'' + z' - 2z = 0, \quad z(t=0) = 4, z'(t=0) = 1$$

Difficulty level   ○ Easy   ● Normal   ○ Hard
Calculation amount   ○ Small   ● Normal   ○ Large

1) $e^t - 3e^{-2t}$

2) $3e^t + e^{-2t}$

3) $e^t + 3e^{-2t}$

4) $3e^t - e^{-2t}$

7.20. Determine the inverse Laplace transform of the following term.

$$F(s) = \frac{1}{s^2+4s+3}$$

Difficulty level   ○ Easy   ● Normal   ○ Hard
Calculation amount   ○ Small   ● Normal   ○ Large

1) $\frac{1}{2}\left(e^{-3t} - e^{-2t}\right)$

2) $\frac{1}{4}\left(e^{-t} - e^{-3t}\right)$

3) $\frac{1}{2}\left(e^{-t} - e^{-3t}\right)$

4) $\frac{1}{4}\left(e^{-3t} - e^{-2t}\right)$

7.21. Solve the differential equation below using Laplace transform.

$$x'' + 5x' + 4x = 3 + 2\delta(t), \quad x(0) = x'(0) = 0$$

Difficulty level   ○ Easy   ● Normal   ○ Hard
Calculation amount   ○ Small   ● Normal   ○ Large

1) $\frac{3}{4} - \frac{1}{3}e^{-t} - \frac{5}{12}e^{-4t}$

2) $\frac{3}{4}te^{-t} - \frac{2}{3}e^{-t} + \frac{1}{3}e^{-4t}$

3) $\frac{3}{4}t - \frac{1}{3}e^{-t} - \frac{5}{12}e^{-4t}$

4) $\delta(t) - 2e^{-t} + \frac{2}{5}e^{-4t}$

7.22. Calculate the inverse Laplace transform of the following term.

$$F(s) = \frac{s^2 + 2s}{\left(s^2 + 2s + 2\right)^2}$$

Difficulty level   ○ Easy   ● Normal   ○ Hard
Calculation amount   ○ Small   ● Normal   ○ Large

1) $te^{-t}\cos t$

2) $te^{-t}\sin t$

3) $te^{t}\cos t$

4) $te^{t}\sin t$

7.23. Calculate the inverse Laplace transform of the following term.

$$F(s) = \frac{s^3 + 2s^2 + 4s + 18}{s^4 + 13s^2 + 36}$$

Difficulty level   ○ Easy   ● Normal   ○ Hard
Calculation amount   ○ Small   ● Normal   ○ Large

1) $\cos 2t + \sin 3t$

2) $\cos 2t - \sin 3t$

3) $\cos 3t + \sin 2t$

4) $\cos 3t - \sin 2t$

7.24. Calculate the inverse Laplace transform of the following term.

$$F(s) = \frac{s - 1}{s^2(s + 1)}$$

Difficulty level   ○ Easy   ● Normal   ○ Hard
Calculation amount   ○ Small   ● Normal   ○ Large

1) $-1 + t + e^{-t}$

2) $1 - t + e^{-t}$

3) $-1 + t + e^{t}$

4) $2 - t - 2e^{-t}$

7.25. Calculate the inverse Laplace transform of the following term.

$$F(s) = \frac{s^2 + 2}{s(s+1)(s+2)}$$

Difficulty level   ○ Easy   ● Normal   ○ Hard
Calculation amount   ○ Small   ○ Normal   ● Large
1) $1 - 3e^t + 3e^{-2t}$
2) $1 - 3e^t - 3e^{-t}$
3) $1 - e^t - 2e^{-t}$
4) $1 - 3e^{-t} + 3e^{-2t}$

7.26. Calculate the value of the definite integral below.

$$\int_0^\infty e^{-3t} \sin 2t \, dt$$

Difficulty level   ○ Easy   ○ Normal   ● Hard
Calculation amount   ● Small   ○ Normal   ○ Large
1) $\frac{2}{13}$
2) $\frac{3}{13}$
3) $\frac{2}{5}$
4) $\frac{3}{5}$

7.27. Determine the inverse Laplace transform of the term below.

$$F(s) = \frac{1}{\sqrt{s}}$$

Difficulty level   ○ Easy   ○ Normal   ● Hard
Calculation amount   ● Small   ○ Normal   ○ Large
1) $\frac{1}{\sqrt{\pi}}$
2) $\alpha \pi t$
3) $\frac{1}{\sqrt{\pi t}}$
4) πt

7.28. Determine the inverse Laplace transform of the term below.

$$F(s) = \text{arc}\left(\cot \frac{s}{\omega}\right)$$

Difficulty level   ○ Easy   ○ Normal   ● Hard
Calculation amount   ● Small   ○ Normal   ○ Large
1) $\frac{1}{t} \sin \omega t$
2) $\frac{1}{t} \cos \omega t$
3) $\frac{1}{t^2} \sin \omega t$
4) $\frac{1}{t^2} \cos \omega t$

7.29. Calculate the value of the definite integral below.

$$\int_0^\infty e^{-3t} \sin t \cos t \, dt$$

Difficulty level   ○ Easy   ○ Normal   ● Hard
Calculation amount   ● Small   ○ Normal   ○ Large

1) $\frac{3}{26}$
2) $\frac{1}{13}$
3) $\frac{3}{14}$
4) $\frac{1}{7}$

7.30. Calculate the value of the definite integral below.

$$\int_0^\infty te^{-4t} \cos 2t \, dt$$

Difficulty level   ○ Easy   ○ Normal   ● Hard
Calculation amount   ● Small   ○ Normal   ○ Large

1) $\frac{3}{100}$
2) $\frac{5}{200}$
3) $\frac{\pi}{100}$
4) $\frac{7}{200}$

7.31. Determine the Laplace transform of the graph shown in Fig. 7.1.

Difficulty level   ○ Easy   ○ Normal   ● Hard
Calculation amount   ● Small   ○ Normal   ○ Large

1) $\frac{1}{s}\left(4e^{-s} - 2e^{-2s} + e^{-3s}\right)$
2) $\frac{1}{s^2}\left(4e^{-s} - 2e^{-2s} + e^{-3s}\right)$
3) $\frac{1}{s}\left(e^{-s} - 2e^{-2s} + e^{-3s}\right)$
4) $\frac{1}{s}\left(4e^{-s} - e^{-2s} + e^{-3s}\right)$

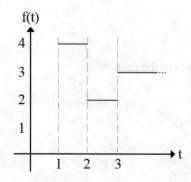

Fig. 7.1 The graph of Problem 7.31

7.32. Determine the Laplace transform of the function below.

$$f(t) = \delta'(t - 2)$$

Difficulty level   ○ Easy   ○ Normal   ● Hard
Calculation amount   ● Small   ○ Normal   ○ Large

1) e^{-2s}
2) $se^{-2s} - 1$
3) se^{-2s}
4) $-2e^{-2s}$

7.33. Calculate the value of the definite integral below.

$$\int_0^\infty e^{-2t} \cos t \, dt$$

Difficulty level   ○ Easy   ○ Normal   ● Hard
Calculation amount   ● Small   ○ Normal   ○ Large

1) $\frac{2}{3}$
2) $\frac{1}{3}$
3) $\frac{2}{5}$
4) $\frac{1}{5}$

7.34. Calculate the inverse Laplace transform of the following term.

$$F(s) = \frac{1}{s^2 + 3s + 4}$$

Difficulty level   ○ Easy   ○ Normal   ● Hard
Calculation amount   ● Small   ○ Normal   ○ Large

1) $e^{-\frac{3}{2}t} \sin \sqrt{7}t$
2) $e^{-\frac{3}{2}t} \sin \frac{\sqrt{7}}{2}t$
3) $\frac{\sqrt{7}}{2} e^{-\frac{3}{2}t} \sin \frac{\sqrt{7}}{2}t$
4) $\frac{2}{\sqrt{7}} e^{-\frac{3}{2}t} \sin \frac{\sqrt{7}}{2}t$

7.35. Calculate the value of the definite integral below.

$$\int_0^\infty e^{-3t} \sin^2 2t \, dt$$

Difficulty level   ○ Easy   ○ Normal   ● Hard
Calculation amount   ● Small   ○ Normal   ○ Large

1) $\frac{3}{75}$
2) $\frac{21}{75}$
3) $\frac{8}{75}$
4) $\frac{12}{49}$

7.36. Calculate the Laplace transform of the graph shown in Fig. 7.2.

Difficulty level   ○ Easy   ○ Normal   ● Hard
Calculation amount   ● Small   ○ Normal   ○ Large

1) $\frac{1}{s}(e^{-4s} - e^{-3s})$
2) $\frac{1}{s}(e^{-4s} + e^{-3s})$
3) $\frac{1}{s}(-e^{-4s} + e^{-3s})$
4) $-\frac{1}{s}(e^{-4s} + e^{-3s})$

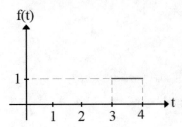

Fig. 7.2 The graph of Problem 7.36

7.37. Solve the integral equation below.

$$f(t) = t + \int_0^t \sin(t - x) f(x) \, dx$$

Difficulty level   ○ Easy   ○ Normal   ● Hard
Calculation amount   ○ Small   ● Normal   ○ Large

1) $f(t) = t + \frac{t^3}{6}$

2) $f(t) = t - \frac{t^3}{6}$

3) $f(t) = \frac{t^2}{2} + \frac{t^4}{12}$

4) $f(t) = \frac{t^2}{2} - \frac{t^4}{12}$

7.38. Calculate the inverse Laplace transform of the following term.

$$F(s) = \frac{3s + 7}{s^2 - 2s - 3}$$

Difficulty level   ○ Easy   ○ Normal   ● Hard
Calculation amount   ○ Small   ● Normal   ○ Large

1) $4e^{3t} - e^t$

2) $4e^{-3t} + e^{-t}$

3) $4e^{-3t} - e^{-t}$

4) $4e^{3t} - e^{-t}$

7.39. Calculate the inverse Laplace transform of the following term.

$$F(s) = \text{arc}(\cot(s + 4))$$

Difficulty level   ○ Easy   ○ Normal   ● Hard
Calculation amount   ○ Small   ● Normal   ○ Large

1) $\frac{e^{-4t}}{t} \sin t$

2) $e^{-4t} \sin t$

3) $\frac{1}{t+4} \sin t$

4) $e^{-4t} \cot t$

7.40. Calculate the inverse Laplace transform of the following term.

$$F(s) = \frac{1}{\sqrt{9s - 1}}$$

Difficulty level   ○ Easy   ○ Normal   ● Hard
Calculation amount   ○ Small   ● Normal   ○ Large

1) $\frac{1}{3\sqrt{t}}e^{\frac{t}{9}}$

2) $\frac{1}{3\sqrt{\pi t}}e^{-\frac{t}{9}}$

3) $\frac{\sqrt{\pi t}}{3}e^{-\frac{t}{9}}$

4) $\frac{1}{3\sqrt{\pi t}}e^{\frac{t}{9}}$

7.41. Solve the integral equation below.

$$y(t) = t^2 + \int_0^t y(t-x)\sin x\,dx$$

Difficulty level   ○ Easy   ○ Normal   ● Hard
Calculation amount   ○ Small   ● Normal   ○ Large

1) $y(t) = t^2 + \frac{t^4}{12}$

2) $y(t) = t^2 + \frac{t^2}{24}$

3) $y(t) = t^2 - \frac{t^4}{12}$

4) $y(t) = t^2(1 + \sin t)$

7.42. Calculate the inverse Laplace transform of the following term.

$$F(s) = s\tan^{-1}\frac{1}{s} - 1$$

Difficulty level   ○ Easy   ○ Normal   ● Hard
Calculation amount   ○ Small   ● Normal   ○ Large

1) $\frac{1}{t^2}(\cos t + \sin t)$

2) $\frac{1}{t^2}(\cos t - \sin t)$

3) $\frac{1}{t^2}(t\cos t + \sin t)$

4) $\frac{1}{t^2}(t\cos t - \sin t)$

7.43. Calculate the value of the following term.

$$f(t) = L^{-1}\left[\frac{4 - 5s}{s^{\frac{3}{2}}}\right]$$

Difficulty level   ○ Easy   ○ Normal   ● Hard
Calculation amount   ○ Small   ● Normal   ○ Large

1) $\frac{7t^{\frac{1}{2}} - 15t^{-\frac{1}{2}}}{\sqrt{\pi}}$

2) $\frac{5t^{\frac{1}{2}} - 8t^{-\frac{1}{2}}}{\sqrt{\pi}}$

3) $\frac{8t^{\frac{1}{2}} - 5t^{-\frac{1}{2}}}{\sqrt{\pi}}$

4) $\frac{15t^{\frac{1}{2}} - 7t^{-\frac{1}{2}}}{\sqrt{\pi}}$

7.44. Calculate the value of the following expression.

$$y(t) = 1 + \int_0^t y(x)\sin(t-x)\,dx$$

Difficulty level   ○ Easy   ○ Normal   ● Hard
Calculation amount   ○ Small   ● Normal   ○ Large

1) $1 + t^2$

2) $t + \frac{t^2}{2}$

3) $1 + \frac{t^2}{2}$

4) $\delta(t) + \frac{t^2}{2}$

7.45. In the equation below, determine the value of $f(t = 0)$.

$$f(t) = L^{-1}\left[\frac{8 - 6s}{16s^2 + 9}\right]$$

Difficulty level   ○ Easy   ○ Normal   ● Hard
Calculation amount   ○ Small   ● Normal   ○ Large

1) $-\frac{3}{8}$

2) 0

3) $\frac{3}{8}$

4) 1

7.46. Determine the Laplace transform of $f(t)$.

$$f(t) = \begin{cases} t & 0 \leq t < 2 \\ 2 & t \geq 2 \end{cases}$$

Difficulty level   ○ Easy   ○ Normal   ● Hard
Calculation amount   ○ Small   ● Normal   ○ Large

1) $\frac{e^{-2s}}{s^2}$

2) $\frac{1 - e^{-2s}}{s}$

3) $\frac{1 + e^{-2s}}{s^2}$

4) $\frac{1 - e^{-2s}}{s^2}$

7.47. Solve the differential equation below using Laplace transform.

$$y'' + y = f(t), \quad y(t = 0) = y'(t = 0) = 0$$

Difficulty level   ○ Easy   ○ Normal   ● Hard
Calculation amount   ● Small   ○ Normal   ○ Large

1) $y(t) = \int_0^t f(x) \cos(t - x)\, dx$

2) $y(t) = \int_0^t f(x)(t - x)^2 dx$

3) $y(t) = \int_0^t f(x)(t - x)\, dx$

4) $y(t) = \int_0^t f(x) \sin(t - x)\, dx$

7.48. Determine the Laplace transform of $f(t)$.

$$f(t) = \begin{cases} 1 & 0 \leq t \leq 2 \\ t & t > 2 \end{cases}$$

Difficulty level   ○ Easy   ○ Normal   ● Hard
Calculation amount   ○ Small   ● Normal   ○ Large

1) $\frac{1}{s} + \frac{1}{s^2}e^{-2s}$

2) $\frac{1}{s} + \frac{1}{s}e^{-2s} + \frac{1}{s^2}e^{-s}$

3) $\frac{1}{s} + \frac{1}{s}e^{-2s} + \frac{1}{s^2}e^{-2s}$

4) $\frac{1}{s} + \frac{1}{s}e^{-s} + \frac{1}{s^2}e^{-2s}$

7.49. Calculate the Laplace transform of the following term.

$$f(t) = \frac{e^t - \cos t}{t}$$

Difficulty level   ○ Easy   ○ Normal   ● Hard
Calculation amount   ○ Small   ● Normal   ○ Large

1) $\ln \frac{s+1}{s-1}$

2) $\ln \frac{\sqrt{s^2+1}}{s-1}$

3) $\ln \frac{s-1}{\sqrt{s^2+1}}$

4) $\ln \frac{s-1}{s+1}$

7.50. Solve the integral equation below.

$$y(t) = 1 + \int_0^t y(x) \sin(t - x)\, dx$$

Difficulty level   ○ Easy   ○ Normal   ● Hard
Calculation amount   ○ Small   ● Normal   ○ Large

1) $y(t) = 1 + \frac{t^2}{2}$

2) $y(t) = \delta(t) + \frac{t^2}{2}$

3) $y(t) = 1 + t^2$

4) $y(t) = t + \frac{t^2}{2}$

7.51. Determine the inverse Laplace transform of the following term.

$$F(s) = \frac{2s^2 - 4}{(s-2)(s+1)(s-3)}$$

Difficulty level   ○ Easy   ○ Normal   ● Hard
Calculation amount   ○ Small   ○ Normal   ● Large

1) $\frac{4}{3}e^{-2t} + \frac{1}{6}e^t - \frac{7}{2}e^{-3t}$

2) $-\frac{4}{3}e^{2t} - \frac{1}{6}e^{-t} + \frac{7}{2}e^{-3t}$

3) $\frac{1}{6}e^{2t} + \frac{3}{4}e^{-t} - \frac{2}{7}e^{-3t}$

4) $-\frac{1}{6}e^{2t} - \frac{3}{4}e^{-t} + \frac{2}{7}e^{3t}$

7.52. Solve the integral equation below.

$$f(t) = \cos bt + c\int_0^t f(t-x)e^{-cx}\, dx$$

Difficulty level   ○ Easy   ○ Normal   ● Hard
Calculation amount   ○ Small   ○ Normal   ● Large

1) $f(t) = c \sin bt + \cos bt$
2) $f(t) = e^{-ct} \cos bt$
3) $f(t) = \frac{c}{b} \sin bt + \cos bt$
4) $f(t) = \frac{c}{b} \sin bt$

7.53. Solve the differential equation below using Laplace transform.

$$y'' - 2y' + y = te^t + 4, \quad y(t = 0) = y'(t = 0) = 1$$

Difficulty level   ○ Easy   ○ Normal   ● Hard
Calculation amount   ○ Small   ○ Normal   ● Large

1) $y(t) = 4te^t - 3e^t + \frac{1}{6}t^3e^t + 4$
2) $y(t) = 3te^t - 4e^t + \frac{1}{6}t^2e^t + 4$
3) $y(t) = 4te^t - 3e^t + \frac{1}{6}te^t + 4$
4) $y(t) = 4te^t - 3e^{-t} + \frac{1}{6}t^3e^t + 4$

7.54. Solve the differential equation below using Laplace transform. Herein, $\delta(t)$ and $u(t)$ are unit Dirac delta function and unit step function, respectively.

$$y'' + 2y' + 2y = \delta\left(t - \frac{\pi}{2}\right), \quad y(x = 0) = y'(x = 0) = 0$$

Difficulty level   ○ Easy   ○ Normal   ● Hard
Calculation amount   ○ Small   ○ Normal   ● Large

1) $-u_{\frac{\pi}{2}}(t)e^{-\left(t-\frac{\pi}{2}\right)} \cos t$
2) $-u(t)e^{-t} \cos t$
3) $u(t)e^{-t} \sin t$
4) $u_{\frac{\pi}{2}}(t)e^{-\left(t-\frac{\pi}{2}\right)} \sin t$

7.55. Solve the following equation.

$$f(x) = L^{-1}\left[\frac{3s^2 - 8s + 3}{s^3 - 4s^2 + 3s}\right]$$

Difficulty level   ○ Easy   ○ Normal   ● Hard
Calculation amount   ○ Small   ○ Normal   ● Large

1) $f(t) = 1 + e^t + e^{3t}$
2) $f(t) = e^t + e^{3t}$
3) $f(t) = -1 - e^t + e^{3t}$
4) $f(t) = 1 - e^t + e^{3t}$

7.56. Calculate the inverse Laplace transform of the following term.

$$F(s) = \frac{s^2 + 2s}{(s^2 + 2s + 2)^2}$$

Difficulty level   ○ Easy   ○ Normal   ● Hard
Calculation amount   ○ Small   ○ Normal   ● Large

1) $te^{-t} \cos t$
2) $te^{-t} \sin t$
3) $te^t \cos t$
4) $te^t \sin t$

7.57. Solve the equation below using Laplace transform.

$$\begin{cases} \dfrac{dx}{dt} + 2\dfrac{dy}{dt} + x - y = 5\sin t \\ \dfrac{dy}{dt} + 3\dfrac{dy}{dt} + x - y = e^t \end{cases}, \quad x(t=0) = y(t=0) = 0$$

Difficulty level   ○ Easy   ○ Normal   ● Hard
Calculation amount   ○ Small   ○ Normal   ● Large

1) $y(t) = -3 + e^t + 2e^{2t} - 5\sin t$
2) $y(t) = -3 + 2e^t + e^{2t} - 5\sin t$
3) $y(t) = -e^t - e^{2t} + 5\sin t + 5\cos t$
4) $y(t) = 3 + e^t + e^{2t} + 5\sin t - 5\cos t$

Solutions of Problems: Laplace Transform and Its Applications in Solving Differential Equations

8

Abstract

In this chapter, the problems of the seventh chapter are fully solved, in detail, step-by-step, and with different methods.

8.1. From Laplace transform table, we have:

$$\frac{1}{s^{n+1}} \overset{L^{-1}}{\Longrightarrow} \frac{t^n}{n!}$$

Therefore, the inverse Laplace transform of the term can be calculated as follows.

$$f(t) = L^{-1}[F(s)] = L^{-1}\left[\frac{1}{s^4}\right] = \frac{t^3}{3!}$$

$$\Rightarrow f(t) = \frac{t^3}{6}$$

Choice (3) is the answer.

8.2. From Laplace transform table, we have:

$$\frac{1}{s^2 + a^2} \overset{L^{-1}}{\Longrightarrow} \frac{1}{a}\sin at$$

Therefore, the inverse Laplace transform of the term is as follows.

$$f(t) = L^{-1}\left[\frac{1}{s^2 + 25}\right] = L^{-1}\left[\frac{1}{s^2 + (5)^2}\right]$$

$$\Rightarrow f(t) = \frac{1}{5}\sin 5t$$

Choice (1) is the answer.

8.3. From Laplace transform table, we have:

$$\frac{1}{s - a} \overset{L^{-1}}{\Longrightarrow} e^{at}$$

Therefore, the inverse Laplace transform of the term is as follows.

$$f(t) = L^{-1}\left[\frac{3}{s+4}\right] = 3L^{-1}\left[\frac{1}{s+4}\right]$$

$$\Rightarrow f(t) = 3e^{-4t}$$

Choice (4) is the answer.

8.4. From Laplace transform table, we have:

$$\frac{1}{s^2 + a^2} \xrightarrow{L^{-1}} \frac{1}{a}\sin at$$

Therefore, the inverse Laplace transform of the term can be calculated as follows.

$$f(t) = L^{-1}\left[\frac{1}{s^2 + 8^2}\right] \Rightarrow f(t) = \frac{1}{8}\sin 8t$$

Choice (4) is the answer.

8.5. From Laplace transform table, we have:

$$\cos at \xrightarrow{L} \frac{s}{s^2 + a^2}$$

Therefore, the Laplace transform of the function is as follows.

$$\Rightarrow L[\cos 2t] = \frac{s}{s^2 + 4}$$

Choice (3) is the answer.

8.6. From Laplace transform table, we have:

$$\sin at \xrightarrow{L} \frac{a}{s^2 + a^2}$$

Therefore, the Laplace transform of the function can be calculated as follows.

$$F(s) = L\left[\frac{\sin 4t}{4}\right] = \frac{1}{4}L[\sin 4t] \Rightarrow F(s) = \frac{1}{4}\frac{4}{s^2 + 16}$$

$$\Rightarrow F(s) = \frac{1}{s^2 + 16}$$

Choice (2) is the answer.

8.7. From Laplace transform table, we have:

$$\sinh at \xrightarrow{L} \frac{a}{s^2 - a^2}$$

The Laplace transform of the function can be calculated as follows.

$$F(s) = L\left[\frac{\sinh 5t}{5}\right] = \frac{1}{5} L[\sinh 5t] = \frac{1}{5} \frac{5}{s^2 - 25}$$

$$\Rightarrow F(s) = \frac{1}{s^2 - 25}$$

Choice (2) is the answer.

8.8. From Laplace transform table, we have:

$$t^n \overset{L}{\Rightarrow} \frac{n!}{s^{n+1}}$$

$$e^{at} \overset{L}{\Rightarrow} \frac{1}{s - a}$$

$$L[e^{at} f(t)] = L[f(t)]|_{s \to s-a} = F(s - a)$$

$$L[tf(t)] = -(L[f(t)])' = -(F(s))'$$

Therefore, the Laplace transform of the function can be determined as follows.

Method 1:

$$F(s) = L[e^t t] = L[t]|_{s \to s-1} = \frac{1}{s^2}|_{s \to s-1} \Rightarrow F(s) = \frac{1}{(s-1)^2}$$

Method 2:

$$F(s) = L[te^t] = -(L[e^t])' = -\left(\frac{1}{s-1}\right)' \Rightarrow F(s) = \frac{1}{(s-1)^2}$$

Choice (4) is the answer.

8.9. From Laplace transform table, we have:

$$\sin at \overset{L}{\Rightarrow} \frac{a}{s^2 + a^2}$$

$$L[e^{at} f(t)] = L[f(t)]|_{s \to s-a} = F(s - a)$$

Therefore, the Laplace transform of the function can be determined as follows.

$$F(s) = L[e^t \sin t] = L[\sin t]|_{s \to s-1} = \frac{1}{s^2 + 1}|_{s \to s-1}$$

$$\Rightarrow F(s) = \frac{1}{(s-1)^2 + 1}$$

Choice (4) is the answer.

8.10. As we know, the convolution of two functions is defined as follows.

$$f(t) * g(t) = \int_0^t f(x)g(t-x)dx$$

Therefore:

$$\Rightarrow f(t) * g(t) = t * 1 = \int_0^t x\,dx = \frac{x^2}{2}\Big|_0^t = \frac{t^2}{2}$$

Choice (4) is the answer.

8.11. From Laplace transform table, we have:

$$\sin at \overset{L}{\Rightarrow} \frac{a}{s^2 + a^2}$$

$$L[e^{at}f(t)] = L[f(t)]|_{s \to s-a} = F(s - a)$$

Therefore:

$$F(s) = L[2e^{-2t}\sin 2t] = 2L[\sin 2t]|_{s \to s+2} = 2\left(\frac{2}{s^2 + 4}\right)\Big|_{s \to s+2}$$

$$\Rightarrow F(s) = \frac{4}{(s+2)^2 + 4}$$

Choice (4) is the answer.

8.12. From Laplace transform table, we have:

$$L[t^2 f] = (L[f])''$$

Therefore:

$$L[f(t)] = L[t^2 \sin t] = (L[\sin t])'' = \left(\frac{1}{s^2 + 1}\right)'' = \left(\frac{-2s}{(s^2 + 1)^2}\right)'$$

$$\Rightarrow L[f(t)] = \frac{6s^2 - 2}{(s^2 + 1)^3}$$

Choice (2) is the answer.

8.13. From Laplace transform table, we have:

$$\frac{1}{s - a} \overset{L^{-1}}{\Longrightarrow} e^{at}$$

Therefore, the inverse Laplace transform of the term can be calculated as follows.

$$f(x) = L^{-1}\left[\frac{1}{(s-2)(s-1)}\right] = L^{-1}\left[\frac{1}{s-2} - \frac{1}{s-1}\right] \Rightarrow f(t) = e^{2t} - e^{t}$$

Choice (3) is the answer.

8.14. From Laplace transform table, we have:

$$\sin at \overset{L}{\Rightarrow} \frac{a}{s^2 + a^2}$$

The Laplace transform of the function can be calculated as follows.

$$F(s) = L[\sin t \cos t] \xrightarrow{\sin t \cos t = \frac{1}{2}\sin 2t} F(s) = L\left[\frac{1}{2}\sin 2t\right] = \frac{1}{2}L[\sin 2t] = \frac{1}{2}\frac{2}{s^2 + 4}$$

$$\Rightarrow F(s) = \frac{1}{s^2 + 4}$$

Choice (3) is the answer.

8.15. From Laplace transform table, we have:

$$\frac{1}{s^{n+1}} \overset{L^{-1}}{\Longrightarrow} \frac{t^n}{n!}$$

$$\frac{1}{s^2 + \omega^2} \overset{L^{-1}}{\Longrightarrow} \frac{1}{\omega}\sin \omega t$$

Therefore, the inverse Laplace transform of the term can be calculated as follows.

$$f(t) = L^{-1}\left[\frac{1}{s^2(s^2 + \omega^2)}\right] = \frac{1}{\omega^2}L^{-1}\left[\frac{1}{s^2} - \frac{1}{s^2 + \omega^2}\right]$$

$$\Rightarrow f(t) = \frac{1}{\omega^2}\left(t - \frac{1}{\omega}\sin \omega t\right)$$

Choice (2) is the answer.

8.16. From Laplace transform table, we have:

$$L[tf(t)] = -(L[f(t)])'$$

$$\sinh at \overset{L}{\Rightarrow} \frac{a}{s^2 - a^2}$$

The Laplace transform of the function can be calculated as follows.

$$F(s) = L\left[\frac{t \sinh 2t}{4}\right] = \frac{1}{4}L[t \sinh 2t]$$

$$\Rightarrow F(s) = \frac{1}{4}\left(-(L[\sinh 2t])'\right) = -\frac{1}{4}\left(\frac{2}{s^2-4}\right)'$$

$$\Rightarrow F(s) = \frac{s}{(s^2-4)^2}$$

Choice (4) is the answer.

8.17. From Laplace transform table, we have:

$$\frac{a}{s^2+a^2} \xRightarrow{L^{-1}} \sin at$$

$$\frac{s}{s^2+a^2} \xRightarrow{L^{-1}} \cos at$$

$$F(s+a) \xRightarrow{L^{-1}} e^{-at}f(t)$$

Therefore, the inverse Laplace transform of the term can be determined as follows.

$$f(t) = L^{-1}\left[\frac{3s+2}{s^2+2s+10}\right] = L^{-1}\left[\frac{3s+3-1}{s^2+2s+1+9}\right] = L^{-1}\left[\frac{3(s+1)-1}{(s+1)^2+9}\right]$$

$$\Rightarrow f(t) = 3L^{-1}\left[\frac{s+1}{(s+1)^2+3^2}\right] - L^{-1}\left[\frac{1}{(s+1)^2+3^2}\right]$$

$$\Rightarrow f(t) = 3e^{-t}\cos 3t - \frac{1}{3}e^{-t}\sin 3t$$

Choice (3) is the answer.

8.18. From Laplace transform table, we have:

$$L^{-1}[F(s-a)] = e^{at}L^{-1}[F(s)]$$

$$\frac{1}{s^2+a^2} \xRightarrow{L^{-1}} \frac{1}{a}\sin at$$

Therefore, the inverse Laplace transform of the term is as follows.

$$f(t) = L^{-1}\left(\frac{1}{s^2+3s+\frac{13}{4}}\right) = L^{-1}\left(\frac{1}{s^2+3s+\frac{9}{4}+1}\right) = L^{-1}\left(\frac{1}{(s+\frac{3}{2})^2+1}\right)$$

$$\Rightarrow f(t) = e^{-\frac{3}{2}t}L^{-1}\left[\frac{1}{s^2+1}\right] = e^{-\frac{3}{2}t}\sin t$$

Choice (2) is the answer.

8.19. Based on the information given in the problem, we have:

$$z'' + z' - 2z = 0$$

$$z(t = 0) = 4, z'(t = 0) = 1$$

From Laplace transform table, we have:

$$L\left[f^{(n)}\right] = s^n L[f] - s^{n-1}f(0) - \cdots - f^{n-1}(0) = s^n F(s) - s^{n-1}f(0) - \cdots - f^{n-1}(0)$$

$$\frac{1}{s-a} \xrightarrow{L^{-1}} e^{at}$$

The equation can be solved using Laplace transform as follows.

$$z'' + z' - 2z = 0 \xrightarrow{L} L[z''] + L[z'] - 2L[z] = 0$$

$$\Rightarrow \left(s^2 Z(s) - sz(0) - z'(0)\right) + (sZ(s) - z(0)) - 2L[z] = 0$$

By applying the primary conditions, we have:

$$\Rightarrow s^2 Z(s) - 4s - 1 + sZ(s) - 4 - 2Z(s) = 0 \Rightarrow \left(s^2 + s - 2\right)Z(s) = 4s + 5$$

$$\Rightarrow Z(s) = \frac{4s+5}{s^2+s-2}$$

$$\xrightarrow{L^{-1}} z(t) = L^{-1}\left[\frac{4s+5}{s^2+s-2}\right] = L^{-1}\left[\frac{4s+5}{(s-1)(s+2)}\right] = L^{-1}\left[\frac{3}{s-1} + \frac{1}{s+2}\right] = 3L^{-1}\left[\frac{1}{s-1}\right] + L^{-1}\left[\frac{1}{s+2}\right]$$

$$\Rightarrow z(t) = 3e^t + e^{-2t}$$

Choice (2) is the answer.

8.20. From Laplace transform table, we have:

$$L^{-1}[F(s-a)] = e^{at}L^{-1}[F(s)]$$

$$\frac{1}{s-a} \xrightarrow{L^{-1}} e^{at}$$

Therefore, the inverse Laplace transform of the term can be calculated as follows.

$$f(t) = L^{-1}\left[\frac{1}{s^2+4s+3}\right] = L^{-1}\left[\frac{1}{(s+2)^2 - 1}\right]$$

$$\Rightarrow f(t) = e^{-2t}L^{-1}\left[\frac{1}{s^2-1}\right] = e^{-2t}L^{-1}\left[\frac{1}{2}\left(\frac{1}{s-1} - \frac{1}{s+1}\right)\right] \Rightarrow f(t) = e^{-2t}\frac{1}{2}(e^t - e^{-t})$$

$$f(t) = \frac{e^{-t} - e^{-3t}}{2}$$

Choice (3) is the answer.

8.21. Based on the information given in the problem, we have:

$$x'' + 5x' + 4x = 3 + 2\delta(t)$$

$$x(0) = x'(0) = 0$$

From Laplace transform table, we have:

$$L\left[f^{(n)}\right] = s^n L[f] - s^{n-1} f(0) - \cdots - f^{(n-1)}(0)$$

$$\delta(t) \overset{L}{\Rightarrow} 1$$

$$\frac{1}{s^{n+1}} \overset{L^{-1}}{\Longrightarrow} \frac{t^n}{n!}$$

$$\frac{1}{s-a} \overset{L^{-1}}{\Longrightarrow} e^{at}$$

The equation can be solved using Laplace transform as follows.

$$x'' + 5x' + 4x = 3 + 2\delta(t) \Rightarrow L[x''] + 5L[x'] + 4L[x] = L[3] + 2L[\delta(t)]$$

$$\overset{L}{\Rightarrow} \left(S^2 X(s) - sx(0) - x'(0)\right) + 5(sX(s) - x(0)) + 4X(s) = \frac{3}{s} + 2$$

Applying the primary conditions:

$$\Rightarrow s^2 X(s) + 5sX(s) + 4X(s) = \frac{3}{s} + 2 \Rightarrow X(s) = \frac{3}{s(s+1)(s+4)} + \frac{2}{(s+1)(s+4)}$$

$$\overset{L^{-1}}{\Longrightarrow} x(t) = L^{-1}[X(s)] = L^{-1}\left[\frac{3}{s(s+1)(s+4)} + \frac{2}{(s+1)(s+4)}\right]$$

$$\Rightarrow x(t) = L^{-1}\left(\frac{\frac{3}{4}}{s} - \frac{1}{s+1} + \frac{\frac{1}{4}}{s+4} + \frac{\frac{2}{3}}{s+1} - \frac{\frac{2}{3}}{s+4}\right)$$

$$\Rightarrow x(t) = \frac{3}{4} - e^{-t} + \frac{1}{4}e^{-4t} + \frac{2}{3}e^{-t} - \frac{2}{3}e^{-4t}$$

$$\Rightarrow x(t) = \frac{3}{4} - \frac{1}{3}e^{-t} - \frac{5}{12}e^{-4t}$$

Choice (1) is the answer.

8.22. From Laplace transform table, we have:

$$L^{-1}[F(s-a)] = e^{at} L^{-1}[F(s)]$$

$$L^{-1}\left[-(F(s))'\right] = tf(t)$$

$$\frac{s}{s^2 + a^2} \overset{L^{-1}}{\Longrightarrow} \cos at$$

Therefore, the inverse Laplace transform of the term can be calculated as follows.

$$f(t) = L^{-1}\left[\frac{s^2 + 2s}{(s^2 + 2s + 2)^2}\right] = L^{-1}\left[\frac{s^2 + 2s + 1 - 1}{(s^2 + 2s + 1 + 1)^2}\right] = L^{-1}\left[\frac{(s + 1)^2 - 1}{((s + 1)^2 + 1)^2}\right]$$

$$\Rightarrow f(t) = e^{-t}L^{-1}\left[\frac{s^2 - 1}{(s^2 + 1)^2}\right] = e^{-t}L^{-1}\left[-\left(\frac{s}{s^2 + 1}\right)'\right]$$

$$\Rightarrow f(t) = e^{-t}t\cos t$$

Choice (1) is the answer.

8.23. From Laplace transform table, we have:

$$\frac{1}{s^2 + a^2} \xrightarrow{L^{-1}} \frac{1}{a}\sin at$$

$$\frac{s}{s^2 + a^2} \xrightarrow{L^{-1}} \cos at$$

Therefore, the inverse Laplace transform of the term can be calculated as follows.

$$f(t) = L^{-1}\left[\frac{s^3 + 2s^2 + 4s + 18}{s^4 + 13s^2 + 36}\right] = L^{-1}\left[\frac{s(s^2 + 4) + 2(s^2 + 9)}{(s^2 + 4)(s^2 + 9)}\right]$$

$$\Rightarrow f(t) = L^{-1}\left[\frac{s}{s^2 + 9} + \frac{2}{s^2 + 4}\right] = L^{-1}\left[\frac{s}{s^2 + 3^2}\right] + L^{-1}\left[\frac{2}{s^2 + 2^2}\right]$$

$$\Rightarrow f(t) = \cos 3t + \sin 2t$$

Choice (3) is the answer.

8.24. From Laplace transform table, we have:

$$\frac{1}{s^{n+1}} \xrightarrow{L} \frac{t^n}{n!}$$

$$\frac{1}{s - a} \xrightarrow{L^{-1}} e^{at}$$

Therefore, the inverse Laplace transform of the term can be calculated as follows.

$$f(t) = L^{-1}\left[\frac{s - 1}{s^2(s + 1)}\right] = L^{-1}\left[\frac{2}{s} - \frac{1}{s^2} - \frac{2}{s + 1}\right] = 2L^{-1}\left[\frac{1}{s}\right] - L^{-1}\left[\frac{1}{s^2}\right] - 2L^{-1}\left[\frac{1}{s + 1}\right]$$

$$\Rightarrow f(t) = 2 - t - 2e^{-t}$$

In the above-written calculations, partial fraction expansion (partial fraction decomposition) technique has been applied on the term of $\frac{s-1}{s^2(s+1)}$ as follows.

$$\frac{s-1}{s^2(s+1)} = \frac{A}{s} + \frac{B}{s^2} + \frac{C}{s+1} = \frac{(A+C)s^2 + (A+B)s + B}{s^2(s+1)}$$

$$\Rightarrow s-1 = (A+C)s^2 + (A+B)s + B \Rightarrow \begin{cases} A+C=0 \\ A+B=1 \\ B=-1 \end{cases} \Rightarrow \begin{cases} C=-2 \\ A=2 \\ B=-1 \end{cases}$$

Choice (4) is the answer.

8.25. From Laplace transform table, we have:

$$\frac{1}{s^{n+1}} \xrightarrow{L^{-1}} \frac{t^n}{n!}$$

$$\frac{1}{s-a} \xrightarrow{L^{-1}} e^{at}$$

The inverse Laplace transform of the term can be calculated as follows.

$$f(t) = L^{-1}[F(s)] = L^{-1}\left[\frac{s^2+2}{s(s+1)(s+2)}\right] = L^{-1}\left[\frac{1}{s} + \frac{-3}{s+1} + \frac{3}{s+2}\right]$$

$$\xrightarrow{L^{-1}} f(t) = 1 - 3e^{-t} + 3e^{-2t}$$

In the above-written calculations, partial fraction expansion (partial fraction decomposition) technique has been applied on the term of $\frac{s^2+2}{s(s+1)(s+2)}$ as follows.

$$\frac{s^2+2}{s(s+1)(s+2)} = \frac{A}{s} + \frac{B}{s+1} + \frac{C}{s+2} = \frac{A(s+1)(s+2) + Bs(s+2) + Cs(s+1)}{s(s+1)(s+2)}$$

$$\frac{s^2+2}{s(s+1)(s+2)} = \frac{(A+B+C)s^2 + (3A+2B+C)s + 2A}{s(s+1)(s+2)}$$

$$\Rightarrow (A+B+C)s^2 + (3A+2B+C)s + 2A = s^2 + 2$$

$$\Rightarrow \begin{cases} A+B+C=1 \\ 3A+2B+C=0 \\ 2A=2 \end{cases} \Rightarrow A=1, B=-3, C=3$$

Choice (4) is the answer.

8.26. Based on the definition of Laplace transform, we have:

$$\int_0^\infty e^{-st} f(t)\, dt = L[f]$$

Therefore:

$$\int_0^\infty e^{-3t} \sin 2t\, dt = L[\sin 2t]\big|_{s=3} = \frac{2}{s^2+4}\bigg|_{s=3} = \frac{2}{13}$$

Choice (1) is the answer.

8.27. From Laplace transform table, we have:

$$\frac{1}{s^{a+1}} \xrightarrow{L^{-1}} \frac{t^a}{\Gamma(\alpha+1)}$$

Therefore, the inverse Laplace transform of the term is as follows.

$$f(t) = L^{-1}\left[\frac{1}{\sqrt{s}}\right] = L^{-1}\left[\frac{1}{s^{1/2}}\right] = \frac{t^{-\frac{1}{2}}}{\Gamma\left(\frac{1}{2}\right)}$$

As we know:

$$\Gamma\left(\frac{1}{2}\right) = \sqrt{\pi}$$

Therefore:

$$\Rightarrow f(t) = \frac{t^{-\frac{1}{2}}}{\sqrt{\pi}} = \frac{1}{\sqrt{\pi t}}$$

Choice (3) is the answer.

8.28. From Laplace transform table, we have:

$$L^{-1}[f] = -\frac{1}{t}L^{-1}[F']$$

$$\sin at \xrightarrow{L} \frac{a}{s^2+a^2}$$

Therefore, the inverse Laplace transform of the term is as follows.

$$f(t) = L^{-1}\left[\text{arccot}\,\frac{s}{\omega}\right] = -\frac{1}{t}L^{-1}\left[\left(\text{arccot}\,\frac{s}{\omega}\right)'\right] = -\frac{1}{t}L^{-1}\left[\frac{-\frac{1}{\omega}}{1+\left(\frac{s}{\omega}\right)^2}\right] = \frac{1}{t}L^{-1}\left[\frac{\omega}{s^2+\omega^2}\right]$$

$$\Rightarrow f(t) = \frac{1}{t}\sin \omega t$$

Choice (1) is the answer.

8.29. Based on the definition of Laplace transform, we have:

$$\int_0^\infty e^{-st}f(t)dt = L[f]$$

Moreover, from Laplace transform table, we have:

$$\sin at \xrightarrow{L} \frac{a}{s^2+a^2}$$

Therefore:

$$\int_0^\infty e^{-3t}\sin t\cos t\,dt = \int_0^\infty e^{-3t}(\sin t\cos t)dt = L[\sin t\cos t]|_{s=3} = \frac{1}{2}L[\sin(2t)]|_{s=3} = \frac{1}{2}\left(\frac{2}{s^2+4}\right)\Big|_{s=3} = \frac{1}{13}$$

Choice (2) is the answer.

8.30. From Laplace transform table, we have:

$$L[tf] = -(L[f])'$$

Based on the definition of Laplace transform, we have:

$$\int_0^\infty e^{-st}f(t)dt = L[f]$$

Therefore:

$$\int_0^\infty te^{-4t}\cos dt = \int_0^\infty e^{-4t}(t\cos 2t)dt = L[t\cos 2t]|_{s=4} = -(L[\cos 2t])'|_{s=4}$$

$$= -\left(\frac{s}{s^2+4}\right)'\Big|_{s=4} = \frac{s^2-4}{(s^2+4)^2}\Big|_{s=4} = \frac{3}{100}$$

Choice (1) is the answer.

8.31. From Laplace transform table, we have:

$$t^n \overset{L}{\Rightarrow} \frac{n!}{s^{n+1}}$$

$$L[u_a(t)f(t-a)] = L[u(t-a)f(t-a)] = e^{-as}L[f(t)] = e^{-as}F(s)$$

The function shown in Fig. 8.1 can be formulated using unit step function as follows.

$$f(t) = 4u_1(t) - 2u_2(t) + u_3(t)$$

$$\overset{L}{\Rightarrow} F(s) = L[f(t)] = 4L[u_1(t)] - 2L[u_2(t)] + L[u_3(t)]$$

$$\Rightarrow F(s) = 4\frac{1}{s}e^{-s} - 2\frac{1}{s}e^{-2s} + \frac{1}{s}e^{-3s}$$

$$\Rightarrow F(s) = \frac{1}{s}\left(4e^{-s} - 2e^{-2s} + e^{-3s}\right)$$

Choice (1) is the answer.

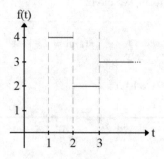

Fig. 8.1 The graph of problem 8.31

8.32. From Laplace transform table, we have:

$$L[f'] = sL[f] - f(0)$$

$$\delta(t-a) = 0 \ for \ t \neq a$$

$$\delta(t-a) \overset{L}{\Rightarrow} e^{-as}$$

Therefore:

$$f(t) = \delta'(t-2) \overset{L}{\Rightarrow} F(s) = L[\delta'(t-2)] \Rightarrow F(s) = sL[\delta(t-2)] - \delta(0-2) \Rightarrow F(s) = se^{-2s} - 0$$

$$\Rightarrow F(s) = se^{-2s}$$

Choice (3) is the answer.

8.33. From Laplace transform table, we have:

$$\int_0^\infty e^{-st}f(t)\,dt = F(s)$$

The definite integral can be solved by using Laplace transform as follows.

$$\int_0^\infty e^{-2t}\cos t\,dt = L[\cos t]|_{s=2} = \frac{s}{s^2+1}|_{s=2} = \frac{2}{5}$$

Choice (3) is the answer.

8.34. From Laplace transform table, we have:

$$L^{-1}[F(s-a)] = e^{at}L^{-1}[F(s)]$$

Therefore, the inverse Laplace transform of the term can be calculated as follows.

$$f(t) = L^{-1}\left[\frac{1}{s^2+3s+4}\right] = L^{-1}\left[\frac{1}{\left(s+\frac{3}{2}\right)^2 + \left(\frac{\sqrt{7}}{3}\right)^2}\right]$$

$$\Rightarrow f(t) = e^{-\frac{3}{2}t}L^{-1}\left[\frac{1}{s^2 + \left(\frac{\sqrt{7}}{2}\right)^2}\right]$$

$$\Rightarrow f(t) = \frac{2}{\sqrt{7}}e^{-\frac{3}{2}t}\sin\frac{\sqrt{7}}{2}t$$

Choice (4) is the answer.

8.35. From Laplace transform table, we have:

$$\int_0^\infty e^{-st}f\,dt = L[f]|_{s=s}$$

$$\frac{1}{s^{n+1}} \overset{L^{-1}}{\Longrightarrow} \frac{t^n}{n!}$$

$$\cos at \overset{L}{\Rightarrow} \frac{s}{s^2 + a^2}$$

The definite integral can be solved by using Laplace transform as follows.

$$I = \int_0^\infty e^{-3t} \sin^2 2t \, dt = L\big[\sin^2 2t\big]\big|_{s=3} = L\left[\frac{1 - \cos 4t}{2}\right]\Big|_{s=3} = \frac{1}{2}\left(\frac{1}{s} - \frac{s}{s^2 + 16}\right)\Big|_{s=3}$$

$$\Rightarrow I = \frac{8}{75}$$

Choice (3) is the answer.

8.36. From Laplace transform table, we have:

$$t^n \overset{L}{\Rightarrow} \frac{n!}{s^{n+1}}$$

$$L[u_a(t)f(t - a)] = L[u(t - a)f(t - a)] = e^{-as}L[f(t)] = e^{-as}F(s)$$

The function shown in Fig. 8.2 can be formulated using unit step function as follows.

$$f(t) = u_3(t) - u_4(t)$$

$$\overset{L}{\Rightarrow} F(s) = L[f(t)] = L[u_3(t) - u_4(t)] = \frac{1}{s}e^{-3s} - \frac{1}{s}e^{-4s}$$

$$F(s) = \frac{1}{s}\left(e^{-3s} - e^{-4s}\right)$$

Choice (3) is the answer.

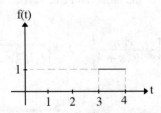

Fig. 8.2 The graph of problem 8.36

8.37. From Laplace transform table, we have:

$$\int_0^t f(x)g(t - x)dx = f(t) * g(t)$$

$$L[f(t) * g(t)] = F(s)G(s)$$

$$\frac{1}{s^{n+1}} \xrightarrow{L^{-1}} \frac{t^n}{n!}$$

$$\sin at \xrightarrow{L} \frac{a}{s^2 + a^2}$$

The equation can be solved using Laplace transform as follows.

$$f(t) = t + \int_0^t \sin(t - x)f(x)dx = t + \sin t * f(t)$$

$$\xrightarrow{L} L[f(t)] = L[t] + L[\sin t * f(t)]$$

$$\Rightarrow L[f(t)] = \frac{1}{s^2} + \frac{1}{1 + s^2} L[f(t)] \Rightarrow \left(1 - \frac{1}{s^2 + 1}\right) L[f(t)] = \frac{1}{s^2}$$

$$\Rightarrow L[f(t)] = \frac{s^2 + 1}{s^4} = \frac{1}{s^2} + \frac{1}{s^4}$$

$$\Rightarrow f(t) = L^{-1}\left[\frac{1}{s^2} + \frac{1}{s^4}\right] = t + \frac{t^3}{6}$$

Choice (1) is the answer.

8.38. From Laplace transform table, we have:

$$L^{-1}[F(s - a)] = e^{at} L^{-1}[F(s)]$$

$$\frac{1}{s - a} \xrightarrow{L^{-1}} e^{at}$$

The inverse Laplace transform of the function can be calculated as follows.

$$f(t) = L^{-1}\left[\frac{3s + 7}{s^2 - 2s - 3}\right] = L^{-1}\left[\frac{3(s - 1) + 10}{(s - 1)^2 - 4}\right]$$

$$\Rightarrow f(t) = e^t L^{-1}\left[\frac{3s + 10}{s^2 - 4}\right] = e^t L^{-1}\left[\frac{4}{s - 2} + \frac{-1}{s + 2}\right] = e^t\left(4e^{2t} - e^{-2t}\right)$$

$$\Rightarrow f(t) = 4e^{3t} - e^{-t}$$

Choice (4) is the answer.

8.39. From Laplace transform table, we have:

$$L^{-1}[F(s)] = -\frac{1}{t} L^{-1}[F'(s)]$$

$$L^{-1}[F(s - a)] = e^{at} L^{-1}[F(s)]$$

$$\frac{1}{s^2 + a^2} \xRightarrow{L^{-1}} \frac{1}{a} \sin at$$

The inverse Laplace transform of the function can be calculated as follows.

$$f(t) = L^{-1}[\text{arc}(\cot(s + 4))]$$

$$\Rightarrow f(t) = -\frac{1}{t} L^{-1} \left[\frac{-1}{1 + (s + 4)^2} \right] = \frac{1}{t} L^{-1} \left[\frac{1}{1 + (s + 4)^2} \right]$$

$$\Rightarrow f(t) = \frac{1}{t} e^{-4t} L^{-1} \left[\frac{1}{1 + s^2} \right]$$

$$\Rightarrow f(t) = \frac{e^{-4t}}{t} \sin t$$

Choice (1) is the answer.

8.40. From Laplace transform table, we have:

$$L^{-1}[F(s - a)] = e^{at} L^{-1}[F(s)]$$

$$\frac{1}{s^{\alpha+1}} \xRightarrow{L^{-1}} \frac{t^{\alpha}}{\Gamma(\alpha + 1)}$$

Therefore, the inverse Laplace transform of the term can be calculated as follows.

$$f(t) = L^{-1} \left[\frac{1}{\sqrt{9s - 1}} \right] = L^{-1} \left[\frac{1}{3\sqrt{s - \frac{1}{9}}} \right]$$

$$\Rightarrow f(t) = \frac{1}{3} e^{\frac{1}{9}t} L^{-1} \left[\frac{1}{\sqrt{s}} \right] = \frac{1}{3} e^{\frac{t}{9}} L^{-1} \left[\frac{1}{s^{\frac{1}{2}}} \right]$$

$$\Rightarrow f(t) = \frac{1}{3} e^{\frac{t}{9}} \frac{t^{-\frac{1}{2}}}{\Gamma\left(\frac{1}{2}\right)}$$

$$\xRightarrow{\Gamma\left(\frac{1}{2}\right) = \sqrt{\pi}} f(t) = \frac{1}{3\sqrt{\pi t}} e^{\frac{t}{9}}$$

Choice (4) is the answer.

8.41. From Laplace transform table, we have:

$$\int_0^t f(t - x) g(x) dx = f(t) * g(t)$$

$$\frac{1}{s^{n+1}} \xRightarrow{L^{-1}} \frac{t^n}{n!}$$

$$L[f(t) * g(t)] = F(s)G(s)$$

The equation can be solved using Laplace transform as follows.

$$y(t) = t^2 + \int_0^t y(t - x) \sin x \, dx \Rightarrow y(t) = t^2 + y(t) * \sin t$$

$$\stackrel{L}{\Rightarrow} Y(s) = \frac{2}{s^3} + \frac{1}{s^2 + 1} Y(s) \Rightarrow Y(s)\left(1 - \frac{1}{s^2 + 1}\right) = \frac{2}{s^3} \Rightarrow Y(s) = \frac{2(s^2 + 1)}{s^5} = 2\left(\frac{1}{s^3} + \frac{1}{s^5}\right)$$

$$\stackrel{L^{-1}}{\Longrightarrow} y(t) = 2L^{-1}\left[\frac{1}{s^3} + \frac{1}{s^5}\right] = 2\left(\frac{t^2}{2} + \frac{t^4}{4!}\right)$$

$$\Rightarrow y(t) = t^2 + \frac{t^4}{12}$$

Choice (1) is the answer.

8.42. From Laplace transform table, we have:

$$L^{-1}[F] = -\frac{1}{t} L^{-1}[F']$$

$$\frac{s}{s^2 + a^2} \stackrel{L^{-1}}{\Longrightarrow} \cos at$$

$$\frac{1}{s^2 + a^2} \stackrel{L^{-1}}{\Longrightarrow} \frac{1}{a} \sin at$$

The inverse Laplace transform of the term can be calculated as follows.

$$f(t) = L^{-1}[F(s)] = L^{-1}\left[s \tan^{-1} \frac{1}{s} - 1\right]$$

$$f(t) = -\frac{1}{t} L^{-1}\left[\left(s \tan^{-1} \frac{1}{s} - 1\right)'\right] = -\frac{1}{t} L^{-1}\left[\tan^{-1} \frac{1}{s} - \frac{s}{s^2 + 1}\right]$$

$$\Rightarrow f(t) = -\frac{1}{t}\left(L^{-1}\left[\tan^{-1} \frac{1}{s}\right] - L^{-1}\left[\frac{s}{s^2 + 1}\right]\right)$$

$$\Rightarrow f(t) = -\frac{1}{t}\left(-\frac{1}{t} L^{-1}\left[\left(\tan^{-1} \frac{1}{s}\right)'\right] - \cos t\right) \Rightarrow f(t) = \frac{1}{t^2}\left(L^{-1}\left[\frac{-1}{s^2 + 1}\right] + t \cos t\right)$$

$$\Rightarrow f(t) = \frac{1}{t^2}(t \cos t - \sin t)$$

Choice (4) is the answer.

8.43. From Laplace transform table, we have:

$$\frac{1}{s^{\alpha+1}} \stackrel{L^{-1}}{\Longrightarrow} \frac{t^\alpha}{\Gamma(\alpha + 1)}$$

$$\Gamma\left(\frac{1}{2}\right) = \sqrt{\pi}$$

$$\Gamma\left(\frac{3}{2}\right) = \frac{\sqrt{\pi}}{2}$$

The inverse Laplace transform of the term can be determined as follows.

$$f(t) = L^{-1}\left[\frac{4 - 5s}{s^{\frac{3}{2}}}\right] = L^{-1}\left[\frac{4}{s^{\frac{3}{2}}} - \frac{5s}{s^{\frac{3}{2}}}\right] = 4L^{-1}\left[\frac{1}{s^{\frac{3}{2}}}\right] - 5L^{-1}\left[\frac{1}{s^{\frac{1}{2}}}\right]$$

$$\Rightarrow f(t) = 4\frac{t^{\frac{1}{2}}}{\Gamma\left(\frac{3}{2}\right)} - 5\frac{t^{-\frac{1}{2}}}{\Gamma\left(\frac{1}{2}\right)} = 4\frac{t^{\frac{1}{2}}}{\frac{\sqrt{\pi}}{2}} - 5\frac{t^{-\frac{1}{2}}}{\sqrt{\pi}}$$

$$f(t) = \frac{8t^{\frac{1}{2}} - 5t^{-\frac{1}{2}}}{\sqrt{\pi}}$$

Choice (3) is the answer.

8.44. From Laplace transform table, we have:

$$\int_0^t f(x)g(t - x)dx = f(t) * g(t)$$

$$L[f(t) * g(t)] = L[f(t)]L[g(t)] = F(s)G(s)$$

$$t^n \overset{L}{\Rightarrow} \frac{n!}{s^{n+1}}$$

$$\sin at \overset{L}{\Rightarrow} \frac{a}{s^2 + a^2}$$

$$\frac{1}{s^{n+1}} \overset{L^{-1}}{\Longrightarrow} \frac{t^n}{n!}$$

We can apply Laplace transform to solve the equation as follows.

$$y(t) = 1 + \int_0^t y(x) \sin(t - x)dx \overset{L}{\Rightarrow} L[y(t)] = L[1] + L\left[\int_0^t y(x) \sin(t - x)dx\right]$$

$$\Rightarrow L[y(t)] = L[1] + L[y(t) * \sin t]$$

$$\Rightarrow Y(s) = \frac{1}{s} + L[y(t)]L[\sin t] \Rightarrow Y(s) = \frac{1}{s} + Y(s)\frac{1}{s^2 + 1} \Rightarrow \left(1 - \frac{1}{s^2 + 1}\right)Y(s) = \frac{1}{s}$$

$$\Rightarrow \frac{s^2}{s^2 + 1}Y(s) = \frac{1}{s} \Rightarrow Y(s) = \frac{1 + s^2}{s^3} = \frac{1}{s^3} + \frac{1}{s}$$

$$\overset{L^{-1}}{\Longrightarrow} y(t) = L^{-1}[Y(s)] = L^{-1}\left[\frac{1}{s^3}\right] + L^{-1}\left[\frac{1}{s}\right]$$

$$\Rightarrow y(t) = \frac{t^2}{2} + 1$$

Choice (3) is the answer.

8.45. From Laplace transform table, we have:

$$\frac{a}{s^2 + a^2} \stackrel{L^{-1}}{\Longrightarrow} \sin at$$

$$\frac{s}{s^2 + a^2} \stackrel{L^{-1}}{\Longrightarrow} \cos at$$

The inverse Laplace transform of the term can be determined as follows.

$$f(t) = L^{-1}\left[\frac{8 - 6s}{16s^2 + 9}\right] = L^{-1}\left[\frac{8 - 6s}{16\left(s^2 + \left(\frac{3}{4}\right)^2\right)}\right] = L^{-1}\left[\frac{1}{2\left(s^2 + \left(\frac{3}{4}\right)^2\right)} - \frac{3s}{8\left(s^2 + \left(\frac{3}{4}\right)^2\right)}\right]$$

$$\Rightarrow f(t) = \frac{1}{2}L^{-1}\left[\frac{1}{s^2 + \left(\frac{3}{4}\right)^2}\right] - \frac{3}{8}L^{-1}\left[\frac{s}{s^2 + \left(\frac{3}{4}\right)^2}\right]$$

$$\Rightarrow f(t) = \frac{1}{2} \times \frac{4}{3}\sin\frac{3}{4}t - \frac{3}{8}\cos\frac{3}{4}t$$

$$\Rightarrow f(0) = -\frac{3}{8}$$

Choice (1) is the answer.

8.46. From Laplace transform table, we have:

$$L[tf] = -(L[f])'$$

$$u_a(t) \stackrel{L}{\Rightarrow} \frac{1}{s}e^{-as}$$

The function of $f(t)$ can be written in terms of unit step function as follows.

$$f(t) = (u_0(t) - u_2(t))t + 2u_2(t)$$

$$\stackrel{L}{\Rightarrow} L[f(t)] = L[(u_0(t) - u_2(t))t] + L[2u_2(t)]$$

$$\Rightarrow F(s) = -(L[u_0(t) - u_2(t)])' + 2L[u_2(t)] \Rightarrow F(s) = -\left(\frac{1}{s} - \frac{1}{s}e^{-2s}\right)' + \frac{2}{s}e^{-2s}$$

$$\Rightarrow F(s) = -\left(-\frac{1}{s^2} + \frac{1}{s^2}e^{-2s} + \frac{2}{s}e^{-2s}\right) + \frac{2}{s}e^{-2s} = \frac{1}{s^2} - \frac{1}{s^2}e^{-2s} - \frac{2}{s}e^{-2s} + \frac{2}{s}e^{-2s}$$

$$\Rightarrow F(s) = \frac{1 - e^{-2s}}{s^2}$$

Choice (4) is the answer.

8.47. Based on the information given in the problem, we have:

$$y'' + y = f(t)$$

$$y(t = 0) = y'(t = 0) = 0$$

From Laplace transform table, we have:

$$L[f''] = s^2 L[f] - sf(0) - f'(0)$$

$$\int_0^t f(x)g(t-x)dx = f(t) * g(t)$$

$$L^{-1}[F(s)G(s)] = L^{-1}[F(s)] * L^{-1}[G(s)]$$

$$L^{-1}\left[\frac{1}{s^2+1}\right] = \sin t$$

The equation can be solved using Laplace transform as follows.

$$y'' + y = f(t) \Rightarrow L[y''] + L[y] = L[f(t)]$$

$$\Rightarrow \left(s^2 Y(s) - sy(0) - y'(0)\right) + Y(s) = F(s)$$

Applying the primary conditions:

$$\Rightarrow s^2 Y(s) + Y(s) = F(s) \Rightarrow \left(s^2 + 1\right)Y(s) = F(s) \Rightarrow Y(s) = F(s)\frac{1}{s^2+1}$$

$$\xrightarrow{L^{-1}} y(t) = L^{-1}[Y(s)] = L^{-1}\left[F(s)\frac{1}{s^2+1}\right]$$

$$\Rightarrow y(t) = f(t) * L^{-1}\left[\frac{1}{s^2+1}\right] = f(t) * \sin t$$

$$\Rightarrow y(t) = \int_0^t f(x)\sin(t-x)dx$$

Choice (4) is the answer.

8.48. From Laplace transform table, we have:

$$u_a(t) \xrightarrow{L} \frac{1}{s}e^{-as}$$

$$L[tf] = -(L[f])'$$

Based on the information given in the problem, we have:

$$f(t) = \begin{cases} 1 & 0 \le t \le 2 \\ t & t > 2 \end{cases}$$

The function of $f(t)$ can be written in terms of unit step function as follows.

$$f(t) = (u_0(t) - u_2(t)) + tu_2(t)$$

$$\overset{L}{\Rightarrow} F(s) = L[f(t)] = L[u_0(t) - u_2(t) + tu_2(t)] = L[u_0(t)] - L[u_2(t)] + L[tu_2(t)]$$

$$\Rightarrow F(s) = \frac{1}{s} - \frac{1}{s}e^{-2s} - \left(\frac{1}{s}e^{-2s}\right)' \Rightarrow F(s) = \frac{1}{s} - \frac{1}{s}e^{-2s} - \left(\frac{-1}{s^2} \times e^{-2s} + \frac{-2}{s} \times e^{-2s}\right)$$

$$\Rightarrow F(s) = \frac{1}{s} + \frac{1}{s}e^{-2s} + \frac{1}{s^2}e^{-2s}$$

Choice (3) is the answer.

8.49. From Laplace transform table, we have:

$$L\left[\frac{1}{t}f(t)\right] = \int_s^\infty L[f(t)]ds$$

$$e^{at} \overset{L}{\Rightarrow} \frac{1}{s-a}$$

$$\cos at \overset{L}{\Rightarrow} \frac{s}{s^2 + a^2}$$

The Laplace transform of the term can be calculated as follows.

$$f(t) = \frac{e^t - \cos t}{t} \overset{L}{\Rightarrow} L[f(t)] = L\left[\frac{e^t - \cos t}{t}\right]$$

$$\Rightarrow F(s) = \int_s^\infty L[e^t - \cos t]ds = \int_s^\infty \left(\frac{1}{s-1} - \frac{s}{s^2+1}\right)ds$$

$$\Rightarrow F(s) = \left(\ln(s-1) - \frac{1}{2}\ln(s^2+1)\right)\Big|_s^\infty = \left(\ln\frac{s-1}{\sqrt{s^2+1}}\right)\Big|_s^\infty = \ln 1 - \ln\frac{s-1}{\sqrt{s^2+1}}$$

$$\Rightarrow F(s) = \ln\frac{\sqrt{s^2+1}}{s-1}$$

Choice (2) is the answer.

8.50. From Laplace transform table, we have:

$$\int_0^t f(x)g(t-x)dx = f(t) * g(t)$$

$$L[f(t) * g(t)] = F(s)G(s)$$

$$\frac{1}{s^{n+1}} \overset{L^{-1}}{\Longrightarrow} \frac{t^n}{n!}$$

The equation can be solved using Laplace transform as follows.

$$y(t) = 1 + \int_0^t y(x) \sin(t-x)dx \Rightarrow y(t) = 1 + y(t) * \sin t$$

$$\overset{L}{\Rightarrow} L[y(t)] = \frac{1}{s} + L[y(t) * \sin t] \Rightarrow Y(s) = \frac{1}{s} + Y(s)\frac{1}{s^2+1}$$

$$\Rightarrow \left(1 - \frac{1}{s^2+1}\right)Y(s) = \frac{1}{s} \Rightarrow \frac{s^2}{s^2+1}Y(s) = \frac{1}{s} \Rightarrow Y(s) = \frac{s^2+1}{s^3} = \frac{1}{s} + \frac{1}{s^3}$$

$$\Rightarrow y(t) = L^{-1}\left[\frac{1}{s} + \frac{1}{s^3}\right] \Rightarrow y(t) = 1 + \frac{t^2}{2}$$

Choice (1) is the answer.

8.51. From Laplace transform table, we have:

$$\frac{1}{s-a} \overset{L^{-1}}{\Longrightarrow} e^{at}$$

First, we need to apply partial fraction expansion (also called partial fraction decomposition) technique on the term as follows.

$$\frac{2s^2-4}{(s-2)(s+1)(s-3)} = \frac{A}{s-2} + \frac{B}{s+1} + \frac{C}{s-3} = \frac{A(s+1)(s-3) + B(s-2)(s-3) + C(s-2)(s+1)}{(s-2)(s+1)(s-3)}$$

$$\Rightarrow \frac{2s^2-4}{(s-2)(s+1)(s-3)} = \frac{(A+B+C)s^2 - (2A+5B+C)s + (-3A+6B-2C)}{(s-2)(s+1)(s-3)}$$

$$\Rightarrow \begin{cases} A+B+C = 2 \\ 2A+5B+C = 0 \\ -3A+6B-2C = -4 \end{cases} \Rightarrow A = -\frac{4}{3}, B = -\frac{1}{6}, C = \frac{7}{2}$$

Now, the inverse Laplace transform of the term can be determined as follows.

$$f(t) = L^{-1}\left[\frac{2s^2-4}{(s-2)(s+1)(s-3)}\right] = L^{-1}\left(\frac{-\frac{4}{3}}{s-2} + \frac{-\frac{1}{6}}{s+1} + \frac{\frac{7}{2}}{s-3}\right)$$

$$\Rightarrow f(t) = -\frac{4}{3}L^{-1}\left[\frac{1}{s-2}\right] - \frac{1}{6}L^{-1}\left[\frac{1}{s+1}\right] + \frac{7}{2}\left[\frac{1}{s-3}\right]$$

$$\Rightarrow f(t) = -\frac{4}{3}e^{2t} - \frac{1}{6}e^{-t} + \frac{7}{2}e^{3t}$$

Choice (2) is the answer.

8.52. From Laplace transform table, we have:

$$\int_0^t f(x)g(t-x)dx = f(t) * g(t)$$

$$L[f(t) * g(t)] = L[f(t)]L[g(t)] = F(s)G(s)$$

$$\cos at \overset{L}{\Rightarrow} \frac{s}{s^2 + a^2}$$

$$\sin at \overset{L}{\Rightarrow} \frac{a}{s^2 + a^2}$$

$$e^{at} \overset{L}{\Rightarrow} \frac{1}{s - a}$$

We can solve the equation using Laplace transform as follows.

$$f(t) = \cos bt + c \int_0^t f(t - x)e^{-cx}dx$$

$$\overset{L}{\Rightarrow} F(s) = L[\cos bt] + cL\left[\int_0^t f(t - x)e^{-cx}dx\right]$$

$$\Rightarrow F(s) = L[\cos bt] + cL[f(t) * e^{-ct}]$$

$$\Rightarrow F(s) = \frac{s}{s^2 + b^2} + cL[f(t)]L[e^{-ct}] = \frac{s}{s^2 + b^2} + cF(s)\frac{1}{s + c}$$

$$\Rightarrow \left(1 - \frac{c}{s + c}\right)F(s) = \frac{s}{s^2 + b^2} \Rightarrow \frac{s}{s + c}F(s) = \frac{s}{s^2 + b^2} \Rightarrow F(s) = \frac{s + c}{s^2 + b^2}$$

$$\overset{L^{-1}}{\Longrightarrow} f(t) = L^{-1}[F(s)] = L^{-1}\left[\frac{s + c}{s^2 + b^2}\right] = L^{-1}\left[\frac{s}{s^2 + b^2}\right] + cL^{-1}\left[\frac{1}{s^2 + b^2}\right]$$

$$\Rightarrow f(t) = \cos bt + \frac{c}{b}\sin bt$$

Choice (3) is the answer.

8.53. Based on the information given in the problem, we have:

$$y'' - 2y' + y = te^t + 4$$

$$y(t = 0) = y'(t = 0) = 1$$

From Laplace transform table, we have:

$$L\left[f^{(n)}\right] = s^n L[f] - s^{n-1}f(0) - \cdots - f^{n-1}(0) = s^n F(s) - s^{n-1}f(0) - \cdots - f^{n-1}(0)$$

$$[e^{at}f(t)] = L[f(t)]|_{s \to s-a}$$

$$L^{-1}[F(s - a)] = e^{at}L^{-1}[F(s)]$$

$$\frac{1}{s^{n+1}} \overset{L^{-1}}{\Longrightarrow} \frac{t^n}{n!}$$

The equation can be solved using Laplace transform as follows.

$$y'' - 2y' + y = te^t + 4 \Rightarrow L[y''] - 2L[y'] + L[y] = L[te^t] + 4L[1]$$

$$\overset{L}{\Rightarrow} (s^2Y - sy(0) - y'(0)) - 2(sY - y(0)) + Y = \frac{1}{(s-1)^2} + \frac{4}{s}$$

Applying the primary conditions:

$$\Rightarrow (s^2Y - s - 1) - (2sY - 2) + Y = \frac{1}{(s-1)^2} + \frac{4}{s}$$

$$\Rightarrow (s^2 - 2s + 1)Y - s + 1 = \frac{1}{(s-1)^2} + \frac{4}{s}$$

$$\Rightarrow (s-1)^2Y(s) = \frac{1}{(s-1)^2} + \frac{4}{s} + (s-1)$$

$$\Rightarrow Y(s) = \frac{1}{(s-1)^4} + \frac{4}{s(s-1)^2} + \frac{1}{s-1}$$

$$\Rightarrow y(t) = L^{-1}[Y(s)] = L^{-1}\left[\frac{1}{(s-1)^4}\right] + 4L^{-1}\left[\frac{1}{s(s-1)^2}\right] + L^{-1}\left[\frac{1}{s-1}\right]$$

$$\Rightarrow y(t) = L^{-1}\left[\frac{1}{(s-1)^4}\right] + 4L^{-1}\left[\frac{1}{s} + \frac{-1}{s-1} + \frac{1}{(s-1)^2}\right] + L^{-1}\left[\frac{1}{s-1}\right]$$

$$\Rightarrow y(t) = e^t L^{-1}\left[\frac{1}{s^4}\right] + 4\left(1 - e^t + e^t L^{-1}\left[\frac{1}{s^2}\right]\right) + e^t$$

$$\Rightarrow y(t) = e^t \frac{t^3}{3!} + 4 - 4e^t + 4e^t t + e^t$$

$$\Rightarrow y(t) = 4te^t - 3e^t + \frac{1}{6}t^3e^t + 4$$

Choice (1) is the answer.

In the above-written calculations, partial fraction expansion (partial fraction decomposition) technique has been applied on the term of $\frac{1}{s(s-1)^2}$ as follows.

$$\frac{1}{s(s-1)^2} = \frac{A}{s} + \frac{B}{s-1} + \frac{C}{(s-1)^2} = \frac{A(s-1)^2 + Bs(s-1) + Cs}{s(s-1)^2} = \frac{s^2(A+B) + s(-2A-B+C) + A}{s(s-1)^2}$$

$$\Rightarrow 1 = s^2(A+B) + s(-2A-B+C) + A \Rightarrow \begin{cases} A = 1 \\ A + B = 0 \\ -2A - B + C = 0 \end{cases} \Rightarrow A = 1, B = -1, C = 1$$

$$\Rightarrow \frac{1}{s(s-1)^2} = \frac{1}{s} - \frac{1}{s-1} + \frac{1}{(s-1)^2}$$

8.54. From Laplace transform table, we have:

$$L\left[f^{(n)}\right] = s^n L[f] - s^{n-1} f(0) - \cdots - f^{(n-1)}(0)$$

$$\delta(t - a) \overset{L}{\Rightarrow} e^{-as}$$

$$L^{-1}\left[e^{-as} F(s)\right] = u_a(t) f(t - a)$$

Based on the information given in the problem, we have:

$$y'' + 2y' + 2y = \delta\left(t - \frac{\pi}{2}\right)$$

$$y(x = 0) = y'(x = 0) = 0$$

The equation can be solved using Laplace transform as follows.

$$y'' + 2y' + 2y = \delta\left(t - \frac{\pi}{2}\right) \Rightarrow L[y''] + 2L[y'] + 2L[y] = L\left[\delta\left(t - \frac{\pi}{2}\right)\right]$$

$$\Rightarrow \left(s^2 Y(s) - sy(0) - y'(0)\right) + 2(sY(s) - y(0)) + 2Y = L\left[\delta\left(t - \frac{\pi}{2}\right)\right]$$

Applying the primary conditions:

$$\Rightarrow s^2 Y(s) + 2sY(s) + 2sY(s) = e^{-\frac{\pi}{2}s} \Rightarrow \left(s^2 + 2s + 2\right) Y(s) = e^{-\frac{\pi}{2}s} \Rightarrow Y(s) = \frac{e^{-\frac{\pi}{2}s}}{(s+1)^2 + 1}$$

$$\overset{L^{-1}}{\Longrightarrow} y(t) = L^{-1}[Y(s)] = L^{-1}\left[\frac{e^{-\frac{\pi}{2}s}}{(s+1)^2 + 1}\right] = u_{\frac{\pi}{2}}(t) L^{-1}\left[\frac{1}{(s+1)^2 + 1}\right]\Bigg|_{t \to t - \frac{\pi}{2}}$$

$$\Rightarrow y(t) = u_{\frac{\pi}{2}}(t)\left(e^{-t} L^{-1}\left[\frac{1}{s^2 + 1}\right]\right)\Big|_{t \to t - \frac{\pi}{2}} = u_{\frac{\pi}{2}}(t)\left(e^{-t} \sin(t)\right)\Big|_{t \to t - \frac{\pi}{2}}$$

$$\Rightarrow y(t) = u_{\frac{\pi}{2}}(t) e^{-\left(t - \frac{\pi}{2}\right)} \sin\left(t - \frac{\pi}{2}\right) \xrightarrow{\sin\left(t - \frac{\pi}{2}\right) = -\cos t} y(t) = -u_{\frac{\pi}{2}}(t) e^{-\left(t - \frac{\pi}{2}\right)} \cos t$$

Choice (1) is the answer.

8.55. From Laplace transform table, we have:

$$\frac{1}{s^{n+1}} \overset{L^{-1}}{\Longrightarrow} \frac{t^n}{n!}$$

$$\frac{1}{s - a} \overset{L^{-1}}{\Longrightarrow} e^{at}$$

Therefore, the inverse Laplace transform of the term can be calculated as follows.

$$f(t) = L^{-1}\left[\frac{3s^2 - 8s + 3}{s^3 - 4s^2 + 3s}\right] = L^{-1}\left[\frac{3s^2 - 8s + 3}{s(s - 3)(s - 1)}\right]$$

$$\overset{L^{-1}}{\Longrightarrow} f(t) = L^{-1}\left[\frac{1}{s} + \frac{1}{s-3} + \frac{1}{s-1}\right] \Rightarrow f(t) = 1 + e^{3t} + e^{t}$$

Choice (1) is the answer.

In the above-written calculations, partial fraction expansion (partial fraction decomposition) technique has been applied on the term of $\frac{3s^2-8s+3}{s(s-3)(s-1)}$ as follows.

$$\frac{3s^2 - 8s + 3}{s^3 - 4s^2 + 3s} = \frac{3s^2 - 8s + 3}{s(s-3)(s-1)} = \frac{A}{s} + \frac{B}{s-3} + \frac{C}{s-1} =$$

$$= \frac{A(s-3)(s-1) + Bs(s-1) + Cs(s-3)}{s(s-3)(s-1)} = \frac{s^2(A+B+C) + s(-4A-B-3C) + (3A)}{s(s-3)(s-1)}$$

$$\Rightarrow 3s^2 - 8s + 3 = s^2(A+B+C) + s(-4A-B-3C) + (3A)$$

$$\Rightarrow \begin{cases} A+B+C = 3 \\ -4A-B-3C = -8 \\ 3A = 3 \end{cases} \Rightarrow A = 1, B = 1, C = 1$$

$$\Rightarrow \frac{3s^2 - 8s + 3}{s^3 - 4s^2 + 3s} = \frac{1}{s} + \frac{1}{s-3} + \frac{1}{s-1}$$

8.56. From Laplace transform table, we have:

$$L^{-1}[F(s-a)] = e^{at}L^{-1}[F(s)]$$

$$L^{-1}[F] = -tL^{-1}\left[\int F ds\right]$$

$$\frac{s}{s^2 + a^2} \overset{L^{-1}}{\Longrightarrow} \cos at$$

Therefore, the inverse Laplace transform of the term can be calculated as follows.

$$f(t) = L^{-1}\left[\frac{s^2 + 2s}{(s^2 + 2s + 2)^2}\right] = L^{-1}\left[\frac{s^2 + 2s + 1 - 1}{(s^2 + 2s + 1 + 1)^2}\right] = L^{-1}\left[\frac{(s+1)^2 - 1}{\left((s+1)^2 + 1\right)^2}\right]$$

$$\Rightarrow f(t) = e^{-t}L^{-1}\left[\frac{s^2 - 1}{(s^2 + 1)^2}\right] = e^{-t}\left(-tL^{-1}\left[\int \frac{s^2 - 1}{(s^2 + 1)^2} ds\right]\right) = e^{-t}\left(-tL^{-1}\left[\frac{-s}{s^2 + 1}\right]\right)$$

$$\Rightarrow f(t) = te^{-t}\cos t$$

In the above-written calculations, $\int \frac{s^2-1}{(s^2+1)^2} ds$ has been solved as follows.

$$I = \int \frac{s^2 - 1}{(s^2 + 1)^2} ds$$

$$\xrightarrow{s \,=\, \tan t \,\Rightarrow\, ds \,=\, \left(1 + \tan^2 t\right) dt} I = \int \frac{\tan^2 t - 1}{\left(\tan^2 t + 1\right)^2} \left(1 + \tan^2 t\right) dt = \int \frac{\tan^2 t - 1}{\tan^2 t + 1} dt = \int \frac{\frac{\sin^2 t}{\cos^2 t} - 1}{\frac{1}{\cos^2 t}} dt$$

$$= \int \frac{\frac{\sin^2 t - \cos^2 t}{\cos^2 t}}{\frac{1}{\cos^2 t}} dt = \int \left(\sin^2 t - \cos^2 t\right) dt = -\int \cos 2t \, dt = -\frac{1}{2} \sin 2t$$

$$\xrightarrow{s \,=\, \tan t \,\Rightarrow\, t \,=\, \mathrm{arc}(\tan s)} I = -\frac{1}{2} \sin 2(\mathrm{arc}(\tan s))$$

$$\xrightarrow{\sin 2a \,=\, \dfrac{2 \tan a}{1 + \tan^2 a}} I = -\frac{1}{2} \frac{2 \tan(\mathrm{arc}(\tan s))}{1 + \tan^2(\mathrm{arc}(\tan s))} = \frac{-s}{1 + s^2}$$

Choice (1) is the answer.

8.57. Based on the information given in the problem, we have:

$$\begin{cases} \dfrac{dx}{dt} + 2\dfrac{dy}{dt} + x - y = 5 \sin t \\[2mm] \dfrac{dy}{dt} + 3\dfrac{dy}{dt} + x - y = e^t \end{cases}$$

$$x(t = 0) = y(t = 0) = 0$$

From Laplace transform table, we have:

$$L[f'(t)] = sF(s) - f(0)$$

$$\sin at \overset{L}{\Rightarrow} \frac{a}{s^2 + a^2}$$

$$e^{at} \overset{L}{\Rightarrow} \frac{1}{s - a}$$

The problem can be solved by using Laplace transform as follows.

$$\begin{cases} \dfrac{dx}{dt} + 2\dfrac{dy}{dt} + x - y = 5 \sin t \\[2mm] \dfrac{dy}{dt} + 3\dfrac{dy}{dt} + x - y = e^t \end{cases} \overset{L}{\Rightarrow} \begin{cases} L\left[\dfrac{dx}{dt}\right] + 2L\left[\dfrac{dy}{dt}\right] + X - Y = \dfrac{5}{s^2 + 1} \\[2mm] L\left[\dfrac{dy}{dt}\right] + 3L\left[\dfrac{dy}{dt}\right] + X - Y = \dfrac{1}{s - 1} \end{cases}$$

$$\Rightarrow \begin{cases} sX + 2sY + X - Y = \dfrac{5}{s^2 + 1} \\[2mm] 2sX + 3sY + X - Y = \dfrac{1}{s - 1} \end{cases} \Rightarrow \begin{cases} (s + 1)X + (2s - 1)Y = \dfrac{5}{s + 1} \\[2mm] (2s + 1)X + (3s - 1)Y = \dfrac{1}{s - 1} \end{cases}$$

$$Y = \frac{\begin{vmatrix} s + 1 & \dfrac{5}{s^2 + 1} \\[2mm] 2s + 1 & \dfrac{1}{s - 1} \end{vmatrix}}{\begin{vmatrix} s + 1 & 2s - 1 \\[2mm] 2s + 1 & 3s - 1 \end{vmatrix}} = -\frac{(s + 1)(s^2 + 1) - 5(2s + 1)(s - 1)}{s(s - 2)(s - 1)(s^2 + 1)} = \frac{5(2s + 1)}{s(s - 2)(s^2 + 1)} - \frac{s + 1}{s(s - 2)(s - 1)}$$

$$\xrightarrow{L^{-1}} y = L^{-1}[Y] = L^{-1}\left[\frac{5(2s+1)}{s(s-2)(s^2+1)} - \frac{s+1}{s(s-2)(s-1)}\right] = L^{-1}\left[\frac{-3}{s} + \frac{1}{s-2} + \frac{2}{s-1} + \frac{-5}{s^2+1}\right]$$

$$\Rightarrow y(t) = -3 + e^{2t} + 2e^t - 5\sin t$$

Choice (2) is the answer.

Index

A
Analytic function, 53

B
Bernoulli differential equation, 19, 20, 22–26
Bessel differential function, 58

C
Characteristic equation of differential equation, 35–41, 43–46,
    48, 53, 54
Complete (exact) differential equation, 13, 14, 18, 23

D
Definite integral, 67–69, 89, 90

F
First order differential equation, 1–27

G
General guess, 40, 43, 46
General solution, 1, 3, 6–8, 29, 31–33, 35–49, 51, 53, 58

H
Homogeneous differential equation, 35–41, 43–46, 48
Homogeneous second-order differential equation
    with the variable coefficients, 53

I
Independent solution, 50
Integrating factor, 3, 5, 13, 14, 18
Inverse Laplace transform, 61, 62, 64–67, 69–71, 73, 74, 77,
    78, 80–87, 89, 91–95, 98, 101, 102
Inverse operator method, 37–43, 45
Irregular singular point, 49, 50, 54, 55

L
Lagrange method, 44
Laplace transform, 61–75, 77–103
Laplace transform table, 77–99, 101–103
Linear differential equation, 9–12, 15–18, 20–25, 27

M
Maclaurin series expansion, 50, 57

N
Non-homogeneous differential equation, 47
Nonsingular point, 53

O
Ordinary point, 53

P
Partial fraction decomposition technique, 85, 86, 98, 100, 102
Partial fraction expansion technique, 85, 86, 98, 100, 102
Particular solutions, 31–33, 37–46
Polynomial of degree n, 41
Primary singular point, 49, 50, 53–56

S
Second-order differential equation, 29–34, 37–46, 53
Separable differential equation, 10, 11, 13–16, 21
Series, 49–51, 53–58

U
Undetermined coefficient method, 40, 43, 46
Unit Dirac delta function, 74
Unit step function, 74, 88, 90, 95, 97

W
Wronskian of functions, 44